Heterogeneous Memory Organizations
in Embedded Systems

Miguel Peón Quirós • Francky Catthoor
José Manuel Mendías Cuadros

Heterogeneous Memory Organizations in Embedded Systems

Placement of Dynamic Data Objects

 Springer

Miguel Peón Quirós
EPFL STI IEL ESL
École Polytechnique Fédérale de Lausanne
Lausanne, Vaud, Switzerland

José Manuel Mendías Cuadros
Facultad de Informática
Universidad Complutense de Madrid
Madrid, Spain

Francky Catthoor
IMEC, Heverlee, Belgium

ISBN 978-3-030-37434-1   ISBN 978-3-030-37432-7   (eBook)
https://doi.org/10.1007/978-3-030-37432-7

This Springer imprint is published by the registered company Springer Nature Switzerland AG.
The registered company address is: Gewerbestrasse 11, 6330 Cham, Switzerland

Preface

Throughout this text, we try to demonstrate the tantalizing possibilities that a careful data placement can offer. We explain a possible methodology for the placement of dynamic data types and apply it on several examples to show exactly how it can be used and to evaluate the possible improvements. We hope that with these examples the readers will be capable of applying those and new techniques to improve the energy efficiency and performance of their designs.

Lausanne, Switzerland   Miguel Peón-Quirós
Heverlee, Belgium   Francky Catthoor
Madrid, Spain   José Manuel Mendías Cuadros

Contents

Acronyms

API	Application program interface
ASIC	Application-specific integrated circuit
AVL	Self-balancing binary search tree invented by Adelson–Velskii and Landis
CNN	Convolutional neural network
CRC	Cyclic redundancy check
DDR	Double data rate
DDT	Dynamic data type
DM	Dynamic memory
DMA	Direct memory access
DMM	Dynamic memory manager
DRAM	Dynamic random access memory
DRR	Deficit round robin
DSP	Digital signal processor
FPB	Frequency of accesses per byte
FPGA	Field programmable gate array
ID	Identifier assigned to a dynamic data type during instrumentation
ILP	Instruction-level parallelism
IP	Internet protocol
JEDEC	JEDEC Solid State Technology Association
LRU	Least recently used
MMU	Memory management unit
NUMA	Non-uniform memory access
NTC	Near-threshold computing
NTV	Near-threshold voltage
NVM	Non-volatile memory
OS	Operating system
QoS	Quality of service
ReRAM	Resistive RAM
SCM	Standard-cell memory
SDRAM	Synchronous dynamic random access memory

SoC	System on a chip
SPM	Scratchpad memory
SRAM	Static random access memory
SSD	Solid-state drive
TCM	Tightly-coupled memory, a synonym for SPM
TCP	Transmission control protocol
WSC	Warehouse-scale computer

Chapter 1
Introduction

The performance of computing systems has experienced an exponential increase over the last decades. However, today computer performance is at a crossroads. The technological and architectural advances that have sustained those performance improvements up to now are increasingly giving diminishing returns. This situation can be traced to three main issues [3]:

- The power wall—or having more transistors than can be powered simultaneously.
- The ILP wall—or diminishing returns for additional hardware.
- The memory wall—or diverging processor and memory speeds.

The power wall owes to the growing difficulties in reducing the energy consumption of transistors on par with their physical dimensions. The power dissipation per area unit is not (almost) constant anymore. As a result, we can now put more transistors in a chip than we can afford to power at once. Therefore, either some of them are turned down at different times, or their operating frequency has to be kept below a limit. Several techniques such as voltage-frequency scaling or aggressive clock gating have been applied to mitigate its effects. The ILP (instruction-level parallelism) wall means that adding more hardware to extract more parallelism from single-threaded code has diminishing results. Therefore, during the last years the focus of interested has moved/displaced to extract the parallelism present at higher levels of abstraction: Thread, process, or request-level parallelism. Finally, the memory wall is due to the fact that processor speeds have increased at a much higher rate than memory access speeds. As a consequence, the execution speed of many applications is dominated by the memory access time.

The three walls challenge together the continuous improvement of the performance of computing systems. However, with time the focus has shifted from the pair ILP-memory walls to the power-memory ones. The reason is that energy has become a major concern in the realms of both large-scale mobile/embedded computing. The memory wall affects system performance, hence putting pressure on mechanisms to improve ILP parallelism, but the memory subsystem also contributes

© Springer Nature Switzerland AG 2020
M. Peón Quirós et al., *Heterogeneous Memory Organizations in Embedded Systems*, https://doi.org/10.1007/978-3-030-37432-7_1

significantly to the total energy consumption of the system. In particular, if the memory subsystem does not perform its work efficiently, for example, executing continuous data movements between memory levels (e.g., "cache trashing"), the total energy consumption of the system will increase because there will be more memory operations and the system will take more time to finish its task, therefore staying longer in a high-power mode. As a result, this work focuses on the (data) memory wall, but keeping an eye on the contribution of the memory subsystem to the total energy consumption.

1.1 The Memory Wall

The most common computational model nowadays is based on processing elements that read both their instructions and data from storage units known as "memories." Soon it was observed that the amount of data that could be stored in the memories was increasing at a faster pace than their access speed. But the real problem was that processor speeds were increasing even faster: Bigger problems could now be solved and hence, more data needed to be accessed. As the discrepancy between processor and data access speeds widened, processors had to stall and wait for the arrival of new instructions and data from the memories more often. This problem, known as the "memory wall," has been one of the main worries of computer architects for several decades. Its importance was outlined by Wulf and McKee [33] in an effort to draw attention and foster innovative approaches.

A fundamental observation is that, given a particular silicon technology, data access speed reduces as memory size increases: Bigger memories have decoders with more logic levels and a larger area means higher propagation delays through longer wires. Additionally, SRAM technology, which has been traditionally employed to build faster memories, requires more area per bit than DRAM. This has the effect that integrating bigger SRAMs increases significantly the chip area, and therefore also its final price.

Those factors led to the idea of combining several memories of varying sizes and access speeds, hence giving birth to the concept of memory hierarchy: A collection of memory modules where the fastest ones are placed close to the processing element and the biggest (and slowest) are placed logically—and commonly also physically—further from it. As suggested by Burks et al. [9], an ideal memory hierarchy would approximate the higher speed of expensive memory technologies that can only be used to build small memories while retaining the lower cost per bit of other technologies that can be used to build larger, although slower, memories.

The presence of memories with different sizes and access characteristics leads inevitably to the following question:

> Which data should be placed into each memory element to obtain the best possible performance?

This question cannot be answered in the same way for static data with relatively high access locality than for dynamic data with low locality. However, as most of the previous literature has studied the former, we will cover that first. Then, we will shift our focus to the dynamic case, which is our main target.

1.1.1 Cache Memories and Explicitly Managed Scratchpads for High-Locality and Static Data

The use of a cache memory, that is, a small and fast memory that holds the most commonly used subset of data or instructions from a larger memory, has been the preferred option to implement the concept of memory hierarchy.[1] With its introduction, computer architecture provided a transparent and immediate mechanism to reduce memory access time by delegating to the hardware the choice of the data objects that should be stored at each level in the memory hierarchy at any given time. The key observation that led to the design of the cache memory is access locality, both temporal and spatial: Data (or instructions) that have been accessed recently have a high probability of being accessed again quite soon, and addresses close to that being accessed at a given moment have a high chance of being accessed next.

In general, cache memories have introduced a significant improvement on data access speeds, but they represent an additional cost in terms of hardware area (both for the storage needed to hold replicated data and for the logic to manage them) and energy consumption. These are substantial issues, especially for embedded systems with constrained energy, temperature, or area budgets. Figure 1.1 summarizes the advantages and consequences of including hardware caches in a computing system.

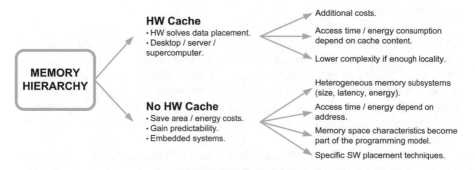

Fig. 1.1 Advantages and disadvantages of hardware-controlled caches

---

[1]A preliminary design for a cache memory was presented by Wilkes [32] under the name of "slave memory." He explained the use of address tags to match the contents of the "lines" in the slave memory with those in the main one and of validity/modification bits to characterize the data held in each line.

The general-purpose logic of cache memories will not be suitable for all types of applications, particularly for those that exhibit low access locality. To overcome that limitation, and/or to avoid their additional hardware costs, several software-controlled mechanisms have been proposed over the years. Those approaches exploit specific knowledge about the applications, in contrast to the automatic, but generic, work of the cache memory. They usually rely on a direct memory access (DMA) module or prefetching instructions in the processor to perform data movements across the elements in the memory hierarchy, which can be introduced implicitly by the compiler or explicitly by the programmer in the form of commands for the DMAs or prefetching instructions for the processor with the goal of bringing data closer to the processor before they are going to be needed, according to the intrinsic properties of each algorithm.

While those techniques may eliminate the overheads introduced by the cache controller logic, they still require a high-locality in the data accesses, especially to achieve significant energy savings. Bringing data that will be accessed just once to a closer memory may reduce the latency of bigger memories. However, each of these movements (a read from a far memory, a write to the closer cache, and a final read from the cache when the processor actually needs the data) consumes some energy. This means a certain waste of energy—and some perhaps harmless waste of bandwidth. These considerations give rise to a new question:

Can we reduce data access latency without sacrificing energy consumption?

Cache memories are an excellent mechanism for the placement of data over several memories with different characteristics that, given the premise of sufficient spatial and temporal locality, have greatly contributed to the continuous increase in computer performance of the last decades alleviating the effects of the memory wall. More importantly, cache memories are completely transparent to the programmer, therefore fulfilling the promise of performance improvements from the side of computer architecture alone.

However, a relevant corpus of existing work [5, 17, 23, 30, 31] has shown that scratchpad memories (small, on-chip static RAMs directly addressable by the software) can be more energy-efficient than caches for static data if a careful analysis of the software applications and their data access patterns is done. With these techniques, the software itself determines explicitly which data must be temporarily copied in the closer memories, usually programming a DMA module to perform the transfer of the next data batch while the processor is working on the current one.

The disadvantage of software-controlled scratchpad memories is that data placement and migration are not performed automatically by the hardware anymore. Instead, the programmer or the compiler must explicitly include instructions to move data across the elements of the memory subsystem. These movements are scheduled according to knowledge extracted during the design phase of the system, usually analyzing the behavior of the applications under typical inputs. This mechanism can provide better performance than generic cache memories, but is costly, platform-dependent, relies on extensive profiling, and requires an agent (compiler or programmer) clever enough to recognize the data access patterns of the application. This approach received significant research effort, especially in academic environments during the early 2000s; one recent industrial example is the Autotiler tool of GreenWaves Technologies [19], which aims at automating the programming of a DMA controller to transfer data between the levels of the memory hierarchy.

Example 1.1 Why do these solutions depend on data locality?

Figure 1.2 illustrates the dependency of caches and scratchpads on temporal locality to amortize the energy cost of data movements over several accesses. First, in (a), the cost of reading a single word from main memory is depicted. In systems with a cache memory, the word will be read from memory and stored in the cache until the processing element accesses it. Even if data forwarding to the processor may save the read from the cache (and hide the access latency), the cost for the write into the cache is ultimately paid. A more complete situation is presented in (b), where a data word is read from main memory, stored in the cache (assuming no forwarding is possible), read from there by the processor, and finally written back, first to the cache and then to the main memory. Energy consumption increases in this case because the cost of writing and reading input data from the cache or scratchpad, and then writing and reading results before posting them to the DRAM, is added to the cost of accessing the data straight from the DRAM. As a result, independently of whether they are performed by a cache controller or a DMA, a net overhead of two writes and two reads to the local memory is created without any reutilization payback. (c) Illustrates a quite unlucky situation that may arise if a word that is frequently accessed is displaced from the cache by another word that is only sporadically used—reducing the likelihood of this situation is the main improvement of associative caches, which are organized in multiple sets. Then, (d) shows how allowing the processor to access specific data objects directly from any element in the memory subsystem can benefit overall system performance: Modifying a single data word consumes the energy strictly needed to perform a read and a write from main memory. Of course, these considerations would have been different if prefetching could have been applied successfully. Finally, (e) presents how a specific data placement can be used to put the most frequently accessed data in a closer memory, while data that are seldom accessed reside in the main memory forcing neither evictions nor further data movements.

The fact that cache memories are not optimal for all situations is also highlighted by Ferdman et al. [15], who show that the tendency to include bigger caches in modern microprocessors can be counterproductive in environments such as warehouse-scale computers (WSCs). In those systems, a myriad of machines work together to solve queries from many users at the same time, exploiting request-level parallelism. The datasets used are frequently so big, and data reuse so low, that caches are continuously overflowed and they just introduce unnecessary delays in the data paths. A better solution will be to reduce the size and energy consumption of the data caches; as a result, more processing elements could be packed together per area unit to increase the number of threads executed concurrently.

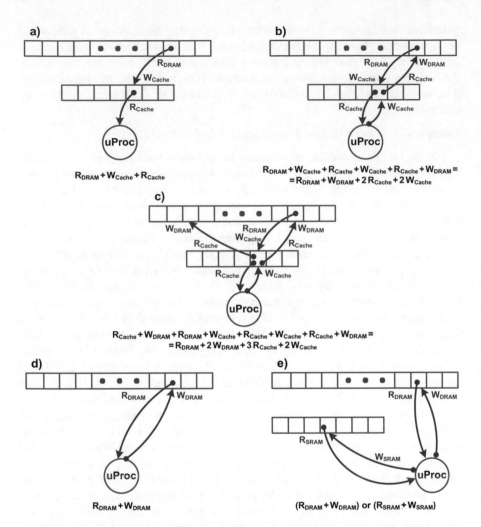

Fig. 1.2 Temporal locality is needed to amortize the cost of data movements across a hierarchical memory subsystem. (**a**) One word is read only once by the processor. (**b**) One word is read and modified by the processor; the result has to be eventually written back to main memory. (**c**) Access to the new word evicts another one previously residing at the same position in the cache. (**d**) Cost of modifying a word if the processing element can access it directly from main memory. (**e**) Cost of modifying data in different levels if the processor can access directly any element of the memory subsystem

1.1.2   The Challenge of Dynamic Data Objects

Both previous mechanisms, caches and software-controlled scratchpads, are efficient for static data objects amenable to prefetching and with a good access locality. However, applications that confront variable cardinality, size, or type of inputs usually rely on dynamic memory and specifically on dynamically linked data structures. Although cache memories can cope with dynamically allocated arrays efficiently, the access pattern to linked structures usually has a devastating effect on access locality, both spatial (thwarting the efforts of prefetching techniques) and temporal (diminishing reuse). Additionally, the dynamic memory mechanism prevents the exact placement in physical memory resources at design time that scratchpad-based techniques require because memory allocation happens at run-time, and the number of instances and their size is potentially unknown until that moment. Although these problems have been known for a long time, most of the previous works on dynamic memory management focused on improving the performance of the dynamic memory managers themselves, reducing the overhead of their internal data structures or palliating the effect of fragmentation, but ignoring the issues derived of placement. Even more important, very little effort has been devoted to the placement of dynamic data objects on heterogeneous memory subsystems.

In summary, caches and scratchpads are not always the most appropriate mechanism for applications that lack a high access locality or that make a heavy use of dynamic memory. We will come back to this in our proposed solution in Chap. 4.

1.2   The Power Wall

While computer architects focused traditionally on improving the performance of computers, energy consumption was a secondary concern solved mainly by techno-logical advances. Quite conveniently, Dennard scaling complemented Moore's law so that, as the number of transistors integrated in a microprocessor increased, their energy efficiency increased as well at an (almost) equivalent rate.

Technological and lithographic advances have enabled an exponential increase on the number of transistors per chip for more than 50 years. This trend was predicted by Gordon Moore in 1965 [26] and revised in 1975 [27]: The number of transistors that could be integrated in a device with the lowest economic cost would double every year (2 years in the revised version). Computer architects exploited these extra transistors (and those also from increasing wafer sizes) adding new capabilities to microprocessors (for instance, exploiting more ILP) with important performance gains. Therefore, performance has actually increased at an even higher rate than transistor density. For personal computers (PCs) alone, performance has doubled every 1.52 years, corresponding to the popular interpretation of Moore's law for increases on performance [25].

In light of these increasing integration capabilities, concerns about heat dissipa-tion appeared early on. However, Moore himself pointed out in his 1965 work that power dissipation density should be constant: *"In fact, shrinking dimensions on an integrated structure makes it possible to operate the structure at higher speed for the same power per unit area."* A few years later, Dennard et al. [12] demonstrated this

proposition for MOSFET devices, in what would be known as "Dennard scaling":
The power dissipation of transistors scales down linearly on par with reductions in
their linear scale. In essence, a scaling factor of $1/k$ in linear dimensions leads to a
reduction of voltage and current of $1/k$, a power dissipation reduction of $1/k^2$ and,
therefore, a constant power density. As the authors point out (on page 265):

> [... T]he power density remains constant. Thus, even if many more circuits are placed on a
> given integrated circuit chip, the cooling problem is essentially unchanged.

Thanks to Dennard scaling, every time the size of transistors was reduced,
more transistors could be integrated into a single circuit while keeping a constant
power requirement. This means that the energy efficiency of the circuits improved
exponentially for several decades, doubling every 1.57 years during the last
60 years [25]. For all this time, computer architects could put to use an exponentially
growing amount of transistors for essentially the same energy budget or, once
reached a minimum performance threshold, design mobile devices with good
enough capabilities and decreasing energy demands.

Dennard scaling held true until about 2004 when it became really difficult to
continue reducing the voltage along with linear dimensions. Many techniques have
been used to palliate the continuous obstacles [8], of which the main one is that
standby power increases significantly as the threshold voltage is reduced. This
problem was hinted by Dennard et al. in their original paper [12] as an issue with
the scalability of subthreshold characteristics and further analyzed in a follow-up 20
years later [11]:

> [...] Therefore, in general, for every 100 mV reduction in V_t the standby current will be
> increased by one order of magnitude. This exponential growth of the standby current tends
> to limit the threshold voltage reduction to about 0.3 V for room temperature operation of
> conventional CMOS circuits.

The consequences of the difficulties to scale down the threshold voltage are
twofold. First, reducing the power-supply voltage brings the electrical level for
the logical "1" closer to that of the logical "0"; hence, it becomes more difficult
to distinguish them reliably. Second, and more importantly, the energy density of
the circuits is not (almost) constant anymore, which means that power dissipation
becomes a much more pressing concern. The situation we face nowadays is
therefore that we can integrate more transistors in a device than what we can
afford to power at the same time [14, 28]. This is the aforementioned "power wall."
Computer architects can no longer focus on improving peak performance relying
on technology improvements to keep energy consumption under control. Novel
techniques are required to limit energy consumption while improving performance
or functionality.

1.2.1   The Advent of Energy-Efficient Computing

As a consequence of the previous events, the first decade of the 2000s witnessed a shift in focus towards obtaining additional reductions in energy consumption. There were two main drivers for this renewed interest on energy efficiency: Mobile computing, supported by batteries, and the increasing concerns about energy consumption and cooling requirements (with their own energy demands) in huge data centers. In this regard, several authors (for example, Koomey et al. [25]) attribute the emergence of mobile computing to the aforementioned exponential increase in energy efficiency during the last 60 years. This effect has enabled the construction of mobile systems supported by batteries and the apparition of a whole new range of applications with increasing demands for ubiquity and autonomy. In essence, computer architects have been able to trade off between improving performance at the same energy level or maintaining a similar performance, which was already good enough in many cases, with lower energy consumption. This property is very useful for battery operated devices, but could also serve as a way to increase energy efficiency in data centers while exploiting request-level parallelism.

Energy consumption is a critical parameter for battery operated devices because optimizations in this area may be directly translated into extended operating times, improved reliability, and lighter batteries. However, in addition to the total amount of energy consumed, the pace at which it is consumed, that is, power, is also relevant. For instance, battery characteristics should be different when a steady but small current needs to be supplied than when bursts of energy consumption alternate with periods of inactivity. Additionally, temporal patterns added to spatial variations in energy consumption may generate "hot spots" in the circuits that contribute to the accelerated aging and reduced reliability of the system [22]. Furthermore, in order to limit system temperature, expensive—and complex—cooling solutions must be added to the design if instantaneous power requirements are not properly optimized.

Apart from these practical issues of operating time and design cost, the ecological and economic aspects of energy play also some role in the domain of embedded systems. Battery operated devices are present in such huge numbers that, even if their individual energy consumption is small, the global effect can still be relevant. On the bright side, every improvement can also potentially benefit vast numbers of units. Therefore, optimizing energy consumption on embedded systems is a fairly critical issue for today's computer architecture.

In a different realm, the shift towards server-side or cloud computing that we are currently experiencing has generated growing concerns about energy consumption in big data centers. Companies such as Google, Amazon, Microsoft, or Facebook have built huge data centers that consume enormous amounts of energy: According to estimates by Van Heddeghem et al. [20], overall worldwide data center energy consumption could have been in the order of 270 TW h in 2012, that is, about 1.4% of the total worldwide electricity production. The sheer amount of machines in those data centers exacerbates any inefficiencies in energy consumption. However, the good news is that all the machines are under the supervision and control of a single entity (i.e., the owning company), so that any improvements can also be applied to huge numbers of machines at the same time.

In this regard, we have seen two interesting trends, with a third one emerging more recently. First, Barroso and Hölzle noticed that most servers were not energy proportional [6]. That is, their energy consumption is not in proportion to their workload; instead, the most energy-efficient operating modes are those with the higher workloads (and operating frequencies). However, as they pointed out, the typical data center server operates usually at low to mid load levels. This important mismatch, which has been the basis for scheduling schemes that try to compact the workload into as few machines as possible, is still reported in subsequent studies such as the ones from Barroso et al. [7] and Arjona et al. [2], although the first one reported that CPU manufacturers (especially Intel) had achieved significant improvements. Second, during the last years, data center architects have speculated with the possibility of favoring simpler architectures. Although these architectures may have lower absolute performance, they typically achieve a higher density of processing elements. This could make them better adapted to exploit request-level parallelism and, most importantly, to increase the energy efficiency of data centers. Finally, in the last decade we have observed that some architectures can operate at near-threshold voltage (NTV) [13, 24], with very interesting trade-offs between energy consumption and operating frequency. Near-threshold computing (NTC) promises a complete turn-around with respect to the previous state of the art: Whereas the previous results suggested that data centers should compact their workloads to improve the energy efficiency of the servers, NTC proposes that the machines should work at lower voltage levels, hence at lower frequencies, and the workloads distributed among more machines [29]. As we can see, energy efficiency is a topic subject to intense research.

Another interesting front is observed in the Internet of Things (IoT) devices, which are connected devices that aim to perform sensing and/or actuating tasks during years without servicing. This is a demanding and evolving realm because, as pointed out by Alioto and Shahghasemi [1], current systems still consume too much energy. In this case, energy consumption has a double impact: First, it determines the lifetime of the device, which may then need replacement or at least a battery change involving (costly) human servicing. Second, the physical size of the required battery limits in many cases the miniaturization of the system. Here, topics such as aggressive use of deep-sleep modes, energy harvesting, pipeline adaptations to varying frequencies [21], or the design of efficient oscillators are relevant topics.

1.3   Energy Consumption in the Memory Subsystem

The memory subsystem is a significant contributor to the overall energy consumption of computing systems, both in server/desktop and mobile platforms. The exact impact varies according to different measurements, platforms, and technologies, but it is usually reported to be between 10% and 30%, with a more pronounced effect in data access dominated applications such as media, networking, or big data processing. For example, Barroso and Hölzle [6] report that the share of DRAM on server peak power² requirements in a Google datacenter is 30%, while in their revised version [7] it is reported as 11.7% (15.2%, excluding the concepts of "power overhead" and "cooling overhead"). Other works, such as the one presented by Carroll and Heiser [10], focus more on mobile platforms.

The problem with the aforementioned studies is that they just differentiate between "CPU power" and "DRAM power." CPUs are composed of processing cores, which contain decoders, control logic, and functional units, but also registers—which can be considered as the first level in the memory subsystem. They also integrate the first, the second, or even the third levels of memory hierarchy in the form of caches (or tightly coupled memories, TCMs). This means that those studies include a significant part of the consumption of the memory subsystem in the concept of "CPU power," therefore lowering the apparent contribution of the memory subsystem to the total energy consumption of the complete system. Unfortunately, getting accurate disclosures of the energy consumed in the different parts of the CPU is not always an easy task. In this work, we will use numbers obtained from SRAM models to calculate the individual contributions of the different memories.

Example 1.2 Memory subsystem impact on energy consumption.

As an example of the impact that different memory subsystem organizations can have on its energy consumption, let us explore here a small subset of the results of the experiment conducted by us with the use of our own simulation framework in Sect. 6.5. The benchmark application uses a trie to create an ordered dictionary of English words, where each node has a list of children directly indexed by letters; it then simulates multiple user look-up operations. This experiment represents a case that is particularly hostile to cache memories because each traversal accesses a single word on each level, the pointer to the next child, but the cache has to move whole lines after every miss. We will explore in more detail the concepts involved in the use of dynamic data structures in Chap. 3.

The performance of the application is evaluated on five different platforms. The reference one (labeled as "Only SDRAM") has just an SDRAM module as main memory. The other platforms add elements to this configuration:

- The platform labeled as "SDRAM, 256 KB Cache" has the SDRAM and a 256 KB cache memory with a line size of 16 words (64 B).
- Platform "SDRAM, 32 KB L1 Cache, 256 KB L2 Cache" has the SDRAM, a first-level 32 KB cache, and a second-level 256 KB cache. Both caches have a line size of 16 words (64 B).

---

²Although in this work we focus about energy consumption, many existing studies present averaged or peak power ratings instead. Unfortunately, the conversion between power and energy is not always straightforward, especially when only peak values are presented.

- Platform "SDRAM, 256 KB Cache (Line size = 4)" has the SDRAM and a 256 KB cache memory with a line size of 4 words (16 B)—in contrast with the previous ones.
- Finally, platform "SDRAM, 256 KB SRAM" has the SDRAM and a 256 KB SRAM memory (also known as a "scratchpad") with an object placement tailored using the techniques presented in this work.

All the caches have an associativity of 16 ways.

Figure 1.3 shows the energy consumption (of the memory subsystem) on each platform, taking the one with only SDRAM as reference. The platform with the SRAM memory ("SDRAM, 256 KB SRAM") is the most efficient, roughly halving the most efficient cache-based platform ("SDRAM, 256 KB Cache (Line size = 4)") and achieving important savings in comparison to the reference case.

The size of the cache line has an important impact on the energy consumption of the system: Because of its longer line size, which is a common feature in modern architectures, the penalty paid by platform "SDRAM, 256 KB Cache" due to the lack of spatial locality in the application is exacerbated up to the point that it would be more efficient to access the SDRAM directly. The use of a multi-level cache hierarchy (platform "SDRAM, 32 KB L1 Cache, 256 KB L2 Cache") does not help in this case; indeed, the additional data movements between the levels in the hierarchy incur an even higher penalty. The reason for this effect appears clearly in Fig. 1.4: The number of accesses actually required by the application (platforms "Only SDRAM" and "SDRAM, 256 KB SRAM") is significantly lower than the number of accesses due to cache-line-wide data movements across elements in the memory hierarchy of the platforms with caches.

Interestingly, Fig. 1.5 shows that, in general, the penalty on the number of cycles spent on the memory subsystem does not increase so dramatically. The reason is

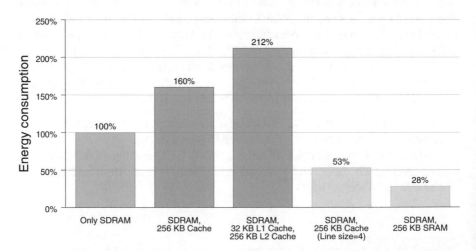

Fig. 1.3 Energy consumption in a DM-intensive benchmark. The results take as reference the energy consumption of the memory subsystem consisting only of an SDRAM module

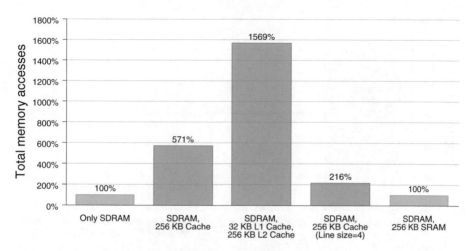

Fig. 1.4 Total number of memory accesses in a DM-intensive benchmark. The results take as reference the number of accesses in the memory subsystem consisting only of an SDRAM module

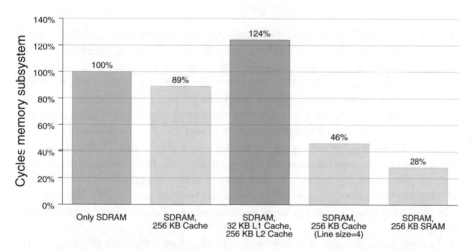

Fig. 1.5 Number of cycles spent on the memory subsystem in a DM-intensive benchmark. The results take as reference the number of cycles on the memory subsystem consisting only of an SDRAM module

that cache memories are much faster than external DRAMs and the latency of some of the superfluous accesses can be hidden through techniques such as pipelining, whereas writing a line to a cache consumes a certain amount of energy regardless of whether all the elements in the line will be used by the processor or not. Energy consumption cannot be "pipelined."

To motivate also for the advantages of combining memories of varying sizes in heterogeneous memory subsystems, Fig. 1.6 presents an additional experiment

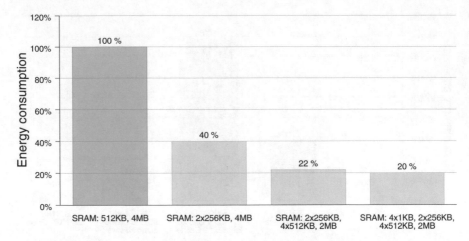

Fig. 1.6 Heterogeneous memory organizations may benefit from the lower energy consumption per access of smaller memories. Also, appropriate placement techniques can place the most accessed data objects on the most efficient memories

with the same application. Now, the reference platform ("SRAM: 512 KB, 4 MB") is configured with enough SRAM capacity to hold all the application dynamic data objects without external DRAMs. This platform has a big module of 4 MB and an additional one of 512 KB—the maximum footprint of the application is in the order of 4.3 MB. Several other configurations that include smaller memories are compared with the previous one. As the energy cost of accessing a word is generally lower for smaller memories, using two 256 KB memories instead of a single 512 KB may reduce energy consumption significantly. Furthermore, with appropriate techniques for dynamic data placement, the most accessed data objects can be placed on the memories that have a lower energy consumption. Thus, heterogeneous memory organizations with partly dynamically allocated data sets can attain important savings in energy consumption with a small increase in design complexity.

In conclusion, the use of data-movement techniques with applications that make an important use of dynamic memory and have low access locality can degrade performance (even if pipelining and other techniques can mitigate this effect to some extent), but, more dramatically, increase energy consumption by futilely accessing and transferring data that are not going to be used. Computer architecture had traditionally focused on improving performance by increasing bandwidth but, especially, reducing latency whereas energy consumption was not so relevant. As the struggle against the power wall becomes more relevant and mobile computing gains relevance, energy consumption is a new goal that deserves specific treatment by computer architects and everybody involved in the design of computing systems.

1.4 Problem Statement and Proposal

In the previous pages we have brought forth the complexities that the use of dynamically allocated objects introduces into the placement problem, especially for generic solutions such as cache memories, but also for more specific solutions created to cope with the placement of static data objects using application-specific knowledge. Additionally, we saw why reducing energy consumption is now as important for computer architecture as improving performance was previously.

In this text, we explore how, for applications that utilize dynamic memory and have low data access locality, using a specifically tailored data placement to avoid or minimize data movements between elements of the memory subsystem brings significant energy savings and performance improvements in comparison with techniques based on data movements such as caches. This goal can be summarized as follows:

> Given a heterogeneous memory subsystem composed of elements with different size, access speed and energy characteristics, and an application that relies on dynamic memory allocation, produce an efficient (especially in terms of energy consumption) placement without data movements of all the dynamically allocated data objects into the memory elements that can be easily implemented at run-time by the operating system (Fig. 1.7).

Our proposal introduces two main questions. First, what is the mechanism that will implement the placement—we propose using the dynamic memory manager itself. Second, what information will be available to make decisions. The interface offered by usual programming languages does not include enough information to allow the dynamic memory manager to produce an efficient data placement. Therefore, we propose to extend that interface with the data type of the objects under creation. In this way, the dynamic memory manager will be able to use prior knowledge about the typical behavior of the instances of the corresponding data type.

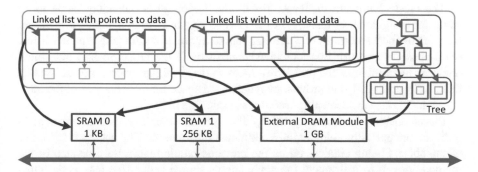

Fig. 1.7 The objective of the methodology presented in this text is to map the dynamic data objects of the application into the elements of the memory subsystem so that the most accessed ones take advantage of the most efficient resources

Avoiding data movements creates a new challenge: If a resource (or part thereof) is reserved exclusively for the instances of one data type, chances are that it will undergo periods of underutilization when only a small number of instances of that data type are alive. To limit the impact of that possibility, we introduce a preliminary analysis step that identifies which data types can be grouped together in the same memory resources, without significant detriment to system performance. Data types can be placed together if their instances are accessed with similar frequency and pattern (so that, for all practical purposes, which instances are placed in a given resource becomes indifferent) or created during different phases of the application execution—hence, not competing for space.

This proposal is compatible with existing techniques for the placement of static data (stack and global objects), such as the ones presented by Kandemir et al. [23], Verma et al. [31] or González-Alberquilla et al. [18], and the normal use of cache memories (through the definition of non-cacheable memory ranges) when those techniques are efficient. It is also adequate for lightweight embedded platforms that contain only small SRAMs instead of a traditional organization with one or several caches backed by a bigger main memory. The designer can use dynamic memory to allocate all the data objects and leave the management of resources and placement considerations to the tool.

1.4.1   Methodology Outline

Figure 1.8 presents the global view of the methodology, whose goal is to produce an explicit placement of dynamic data objects on the memory subsystem. Data movements between memory modules are avoided because they are effective only in situations of high data access locality—moving data that are going to be used just once is not very useful. Thus, the advantages of the methodology increase when access locality is low, as is often the case when dynamic memory is heavily used.

The most significant challenge for the methodology is balancing between an exclusive assignment of resources to DDTs and keeping a high resource utilization: Keeping the best resources idle will not help to improve system performance. Thus, we propose to perform first an extensive analysis of the DDTs to find out which ones can be combined and assign resources to the resulting groups, not to isolated DDTs. Grouping helps in provisioning dedicated space for the instances of the most accessed DDTs, while keeping resource exploitation high.

Placement can be implemented by the dynamic memory manager which, to perform that extended role, requires additional information: The data type identifier of the objects being created. Given the previous requirements, the first step of the methodology is to instrument the application's source code. This task is designed to reduce overall effort; hence, it serves both to conduct an extensive profiling of the application at design time and to supply the dynamic memory manager with the required additional information at run-time. Most of the burden of manual intervention on the source code involves subclassing the declarations of DDTs to

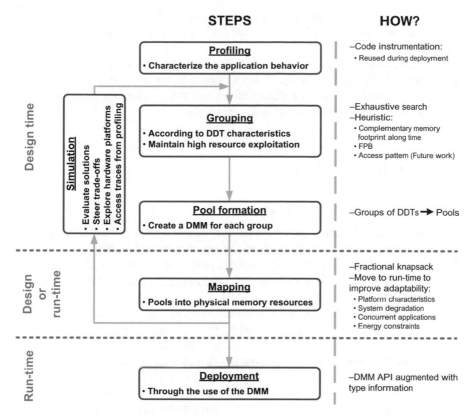

Fig. 1.8 For applications that use dynamic memory and with low access locality, we propose a methodology that groups data objects according to their characteristics and places them accordingly on the platform's memory resources, reducing data movements across the memory subsystem

extract during profiling the number of objects created and destroyed, their final addresses (to relate the corresponding memory accesses) and the type of each object created or destroyed, which is the part of the instrumentation that remains at runtime so that the DMM can implement the placement decisions.

The methodology divides the placement process in two steps: First, the DDTs are grouped according to their characteristics; then, the groups (not the individual DDTs) are placed into the memory resources. The heaviest parts are the grouping of DDTs and the construction of the DMMs themselves. These processes are performed at design time. The work of mapping every group on the platform's memory resources is a simpler process that can be executed at design time or delayed until run-time. The main benefit of delaying the mapping step is the ability to produce a placement suited for the resources actually available in the system at run-time. For example, the application could be efficiently executed (without recompilation) on any instantiation of the platform along a product line, or it could adapt itself to yield a graceful degradation of system performance as the device ages and some resources start to fail—more critically, this could help to improve device reliability.

As explained in Chap. 4, placement is a complex problem and there is no guarantee that the approach adopted in this methodology is optimal. Therefore, we include an additional step of simulation to evaluate the solutions generated and offer to the designer the possibility of steering trade-offs.

1.4.2 Methodology Steps

The methodology is divided in seven steps:

1. **Instrumentation.** The source code of the application is instrumented to generate a log file that contains memory allocation and data access events.
2. **Profiling and analysis.** The application is profiled under typical input scenarios, producing the log file that will be one of the inputs for the tool that implements the methodology.
3. **Group creation.** The number of alive instances of a dynamic data type (DDT), that is, the number of objects created by the application and not yet destroyed, varies along time in accordance to the structure and different phases of the application. The total footprint of the DDTs varies along time accordingly. As a consequence, if each DDT were allocated a memory area in exclusivity, and given that no data movements are executed at run-time, precious memory resources would remain underexploited during relevant fractions of the execution time.

   To tackle that problem, the grouping step clusters the DDTs according to their characteristics (memory footprint evolution along time, frequency of accesses per byte, etc.) as it explores the trade-off between creating a pool for each DDT and merging them all in a single pool. Two important concepts introduced in this step are liveness and exploitation ratio.

   The process of grouping also helps to simplify the subsequent construction of efficient dynamic memory managers because each of them will have to support a smaller set of DDTs and, hence, specific behaviors.
4. **Definition of pool algorithms.** Each pool has a list of DDTs, the required amount of memory for the combined—not added—footprint of the DDTs that it will contain, and the description of the algorithms and internal data structures that will be employed for its management. The appropriate algorithm and organization for each pool can be selected using existing techniques for the design of dynamic memory managers such as the ones presented by Atienza et al. [4]. The DMMs are designed as if each pool were the only one present in the application, which greatly simplifies the process.
5. **Mapping into memory resources.** The description of the memory subsystem is used to map all the pools into the memory modules of the platform, assigning physical addresses to each pool. This step is an instance of the classic fractional

knapsack problem; hence, it can be solved with a greedy algorithm in polynomial time. The output of this step, which is the output of the whole methodology, is the list of memory resources where each of the pools is to be placed.

6. **Simulation and evaluation.** The methodology includes a simulation step based on the access traces obtained during profiling to evaluate the mapping solutions before deployment into the final platform and adjust the values of the parameters that steer the various trade-offs. Additionally, if the exploration is performed at an early design stage when the platform is still unfinished, the results of the simulation can be used to steer the design or selection of the platform in accordance to the needs of the applications.

7. **Deployment.** Finally, the description of the pools needed to perform the allocation and placement of the application DDTs is generated as metadata that will be distributed with the application. The size of the metadata should not constitute a considerable overhead on the size of the deployed application. In order to attain maximum flexibility, a factory of DM managers and the strategy design pattern [16] can be employed at run-time to construct the required memory managers according to their description.

1.5 Text Organization

The rest of this text is organized as follows (Fig. 1.9). Chapter 2 surveys briefly on related works and previous state of the art. Then, Chap. 3 presents an introduction to the main concepts of dynamic memory and dynamic data structures that may aid the unaccustomed reader in understanding the rest of this work. It also explores how the specific access patterns of dynamic data structures can affect the data locality of the application. In Chap. 4, we describe the methodology for dynamic data placement that forms the core of this text. First, we analyze briefly the design space for the construction of dynamic memory managers capable of implementing data placement. Due to the complexity of the problem, we present the simpler approach,

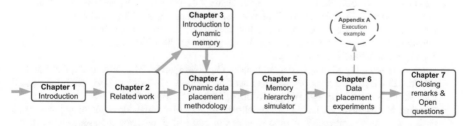

Fig. 1.9 Organization of this text. After the introduction of Chaps. 1, 4, 5, and 6 present our proposal for dynamic data placement and evaluate it. Appendix A explains how the tool that implements our methodology is applied in a concrete case. Finally, Chap. 7 draws closing remarks and outlines the most important open questions that we face nowadays

consisting of a grouping and a mapping step, that we continue using through this work to tackle it.

The rest of the chapter explains thoroughly each of the steps in the methodology, presenting the corresponding algorithms and the parameters available for the designer to steer their work.

Chapter 5 describes in deep detail the implementation of the simulator that we use to evaluate the solutions produced by our methodology (and maybe to modify some algorithm parameters) or to explore different platform design options.

In Chap. 6, we present the results obtained after applying the methodology in a set of case studies. Each of these easily understandable examples is explored under many different combinations of platform resources. With them, we not only show the improvements that can be attained, but we also try to explain the reasons behind those improvements. This chapter includes as well an extensive discussion on the properties, applicability, and drawbacks of the methodology; the conditions under which our experiments were performed and some directions for further improvement.

In Chap. 7, we draw our closing remarks and present promising directions for future work on data placement. We also try to motivate the importance of data placement in environments other than embedded systems, specifically in big data centers, under the light of recent technological developments such as non-volatile memories (NVMs). Finally, Appendix A explains how $\mathcal{D}yn\mathcal{A}s\mathcal{T}$, the tool that implements our methodology, is applied on a concrete case.

References

1. Alioto, M., Shahghasemi, M.: The internet of things on its edge: trends toward its tipping point. IEEE Consum. Electron. Mag. **7**(1), 77–87 (2018). https://doi.org/10.1109/MCE.2017.2755218
2. Arjona Aroca, J., Chatzipapas, A., Fernández Anta, A., Mancuso, V.: A measurement-based analysis of the energy consumption of data center servers. In: Proceedings of the International Conference on Future Energy Systems (e-Energy), pp. 63–74. ACM Press, Cambridge (2014). https://doi.org/10.1145/2602044.2602061
3. Asanovic, K., Bodik, R., Catanzaro, B.C., Gebis, J.J., Husbands, P., Keutzer, K., Patterson, D.A., Plishker, W.L., Shalf, J., Williams, S.W., Yelick, K.A.: The landscape of parallel computing research: a view from Berkeley. Tech. rep., Electrical Engineering and Computer Sciences, University of California at Berkeley (2006). http://www.eecs.berkeley.edu/Pubs/TechRpts/2006/EECS-2006-183.pdf
4. Atienza Alonso, D., Mamagkakis, S., Poucet, C., Peón-Quirós, M., Bartzas, A., Catthoor, F., Soudris, D.: Dynamic Memory Management for Embedded Systems. Springer International Publishing, Switzerland (2015). https://doi.org/10.1007/978-3-319-10572-7
5. Banakar, R., Steinke, S., Lee, B.S., Balakrishnan, M., Marwedel, P.: Scratchpad memory: a design alternative for cache on-chip memory in embedded systems. In: Proceedings of the International Symposium on Hardware/Software Codesign (CODES), pp. 73–78. ACM Press, Estes Park (2002). https://doi.org/10.1145/774789.774805
6. Barroso, L.A., Hölzle, U.: The case for energy-proportional computing. Computer **40**(12), 33–37 (2007). https://doi.org/10.1109/MC.2007.443

7. Barroso, L.A., Clidaras, J., Hölzle, U.: The Datacenter as a Computer: An Introduction to the Design of Warehouse-Scale Machines, 2nd edn. Synthesis Lectures on Computer Architecture, vol. 8(3), pp. 1–154. Morgan and Claypool (2013). https://doi.org/10.2200/S00516ED2V01Y201306CAC024

8. Bohr, M.: A 30 year retrospective on Dennard's MOSFET scaling paper. IEEE Solid-State Circuits Soc. Newsletter 12(1), 11–13 (2007). https://doi.org/10.1109/N-SSC.2007.4785534

9. Burks, A.W., Goldstine, H.H., Neumann, J.: Preliminary discussion of the logical design of an electronic computing instrument, pp. 92–119 (1946). http://www.cs.princeton.edu/courses/archive/fall10/cos375/Burks.pdf

10. Carroll, A., Heiser, G.: An analysis of power consumption in a smartphone. In: Proceedings of the USENIX Annual Technical Conference (USENIX ATC), pp. 21–21. USENIX Association, Boston (2010). http://dl.acm.org/citation.cfm?id=1855840.1855861

11. Davari, B., Dennard, R.H., Shahidi, G.G.: CMOS scaling for high performance and low power—The next ten years. Proc. IEEE 83(4), 595–606 (1995). https://doi.org/10.1109/5.371968

12. Dennard, R.H., Gaensslen, F.H., Rideout, V.L., Bassous, E.: Design of ion-implanted MOS-FET's with very small physical dimensions. IEEE J. Solid State Circuits 9(5), 256–268 (1974). https://doi.org/10.1109/JSSC.1974.1050511

13. Dreslinski, R.G., Wieckowski, M., Blaauw, D., Sylvester, D., Mudge, T.: Near-threshold computing: reclaiming Moore's law through energy efficient integrated circuits. Proc. IEEE 98(2), 253–266 (2010). https://doi.org/10.1109/JPROC.2009.2034764

14. Esmaeilzadeh, H., Blem, E., St. Amant, R., Sankaralingam, K., Burger, D.: Dark silicon and the end of multicore scaling. ACM SIGARCH Comput. Archit. News 39(3), 12 (2011). https://doi.org/10.1145/2024723.2000108

15. Ferdman, M., Adileh, A., Kocberber, O., Volos, S., Alisafaee, M., Jevdjic, D., Kaynak, C., Popescu, A.D., Ailamaki, A., Falsafi, B.: Clearing the clouds: a study of emerging scale-out workloads on modern hardware. In: International Conference on Architectural Support for Programming Languages and Operating Systems (ASPLOS), pp. 37–48. ACM Press, London (2012). https://doi.org/10.1145/2150976.2150982

16. Gamma, E., Helm, R., Johnson, R., Vlissides, J.: Design Patterns: Elements of Reusable Object-Oriented Software. Addison-Wesley Longman Publishing, Boston (1995)

17. Geelen, B., Brockmeyer, E., Durinck, B., Lafruit, G., Lauwereins, R.: Alleviating memory bottlenecks by software-controlled data transfers in a data-parallel wavelet transform on a multicore DSP. In: Proceedings of the IEEE BENELUX/DSP Valley Signal Processing Symposium (SPS-DARTS), pp. 143–146 (2005)

18. González-Alberquilla, R., Castro, F., Piñuel, L., Tirado, F.: Stack filter: reducing L1 data cache power consumption. J. Syst. Archit. 56(12), 685–695 (2010). https://doi.org/10.1016/j.sysarc.2010.10.002

19. GreenWaves Technologies: GAP8 Auto-tiler manual. https://greenwaves-technologies.com/manuals/BUILD/AUTOTILER/html/index.html. Accessed May 2019

20. Heddeghem, W.V., Lambert, S., Lannoo, B., Colle, D., Pickavet, M., Demeester, P.: Trends in worldwide ICT electricity consumption from 2007 to 2012. Comput. Commun. 50, 64–76 (2014). https://doi.org/10.1016/j.comcom.2014.02.008

21. Jain, S., Lin, L., Alioto, M.: Dynamically adaptable pipeline for energy-efficient microarchitectures under wide voltage scaling. IEEE J. Solid State Circuits 53(2), 632–641 (2018). https://doi.org/10.1109/JSSC.2017.2768406

22. JEDEC: Failure Mechanisms and Models for Semiconductor Devices—JEP122G. JEDEC Solid State Technology Association, Arlington (2011)

23. Kandemir, M., Kadayif, I., Choudhary, A., Ramanujam, J., Kolcu, I.: Compiler-directed scratchpad memory optimization for embedded multiprocessors. IEEE Trans. Very Large Scale Integr. Syst. 12, 281–287 (2004)

24. Khare, S., Jain, S.: Prospects of near-threshold voltage design for green computing. In: International Conference on VLSI Design, pp. 120–124 (2013). https://doi.org/10.1109/VLSID.2013.174

25. Koomey, J.G., Berard, S., Sanchez, M., Wong, H.: Implications of historical trends in the electrical efficiency of computing. IEEE Ann. Hist. Comput. **33**(3), 46–54 (2011). https://doi. org/10.1109/MAHC.2010.28
26. Moore, G.E.: Cramming more components onto integrated circuits. Electronics **38**(8), 4 (1965). http://www.cs.utexas.edu/~pingali/CS395T/2013fa/papers/moorespaper.pdf
27. Moore, G.E.: Progress in digital integrated electronics. In: International Electron Devices Meeting, vol. 1, pp. 11–13 (1975)
28. Mudge, T., Hölzle, U.: Challenges and opportunities for extremely energy-efficient processors. IEEE Micro **30**(4), 20–24 (2010). http://ieeexplore.ieee.org/xpl/articleDetails.jsp?arnumber= 5550996
29. Pahlevan, A., Qureshi, Y.M., Zapater Sancho, M., Bartolini, A., Rossi, D., Benini, L., Atienza Alonso, D.: Energy proportionality in near-threshold computing servers and cloud data centers: consolidating or not? In: Proceedings of Design, Automation and Test in Europe (DATE), pp. 147–152 (2018)
30. Panda, P.R., Dutt, N.D., Nicolau, A.: On-chip vs. off-chip memory: the data partitioning problem in embedded processor-based systems. ACM Trans. Design Autom. Electron. Syst. **5**(3), 682–704 (2000). https://doi.org/10.1145/348019.348570
31. Verma, M ., Wehmeyer, L., Marwedel, P.: Cache-aware scratchpad allocation algorithm. In: Proceedings of Design, Automation and Test in Europe (DATE) (2004)
32. Wilkes, M.V.: Slave memories and dynamic storage allocation. IEEE Trans. Electron. Comput. **EC-14**(2), 270–271 (1965). https://doi.org/10.1109/PGEC.1965.264263
33. Wulf, W.A., McKee, S.A.: Hitting the memory wall: implications of the obvious. ACM SIGARCH Comput. Archit. News **23**(1), 20–24 (1995). https://doi.org/10.1145/216585. 216588

Chapter 2
Related Work

2.1 Code Transformations to Improve Access Locality

Memory hierarchies with cache memories are useful to narrow the gap between the speed of processors and that of memories. However, they are only useful if the algorithms generate enough data access locality. Thus, for many years researchers have used techniques that modify the layout of (static) data objects or the scheduling of instructions in code loops to increase access locality. Examples of those techniques are loop transformations such as tiling (blocking) that enhance caching of arrays [2, 17, 35], or the data transfer and storage exploration [13] methodology, which is one of the several proposals to tackle code scheduling at different levels of abstraction (i.e., from the instruction to the task levels), targeting either energy or performance improvements.

2.2 SW-Controlled Data Layout and Static Data Placement

The previous techniques aim to improve data access locality, particularly for systems that include cache memories. However, many works showed that scratchpad memories (small, on-chip SRAM memories directly addressable by the software) can be more energy-efficient than caches for static data if a careful analysis of the applications and their data access patterns is done [24, 31, 42, 54]; among them, the work of Banakar et al. [8] is particularly descriptive. Therefore, two main groups of works were conducted to explore the use of scratchpad memories: Those ones that use data movements to implementing a kind of software caching and prefetching, and those others that produce a fixed data placement for statically allocated data such as global variables and the stack.

The fundamental idea of the works in the first group is to exploit design time knowledge about the application so that the software itself determines explicitly

© Springer Nature Switzerland AG 2020
M. Peón Quirós et al., *Heterogeneous Memory Organizations*
in Embedded Systems, https://doi.org/10.1007/978-3-030-37432-7_2

which data needs to be temporarily copied to the closer memories, producing a dynamic data layout that is usually implemented programming a direct memory access (DMA) controller to copy blocks of data between main memory and the scratchpad while the processor works on a different data block [1, 19, 24, 58]. As with hardware caches, those methods are only useful if the algorithms present enough data access locality [58]. Otherwise, performance may still be improved via prefetching, but energy consumption increases.[1] Ultra-low power embedded devices are examples of architectures that integrate scratchpad or tightly coupled memories (TCMs) in their data path [22, 46]. To alleviate the burden of managing the required data transfers, in the last years commercial companies have started to offer tools that assist the programmer in the process of instrumenting their code to implement double buffering for static data objects between the different levels of the memory hierarchy via a DMA module [26].

The works in the second group aimed to statically assign space in the scratchpad to the most accessed data (or code) objects in the application [30, 31, 50, 53, 54]. Panda et al. [42] and Benini and de Micheli [10] presented good overviews of several techniques to map stack and global variables. Regarding specifically the stack, a hardware structure to transparently map it into a scratchpad memory is presented by González-Alberquilla et al. [25]. Soto et al. [49] explore, both from the perspective of an exact solver and using heuristics, the problem of placing (static) data structures in a memory subsystem where several memories can be accessed in parallel. Their work can be partially seen as a generalization of the mapping step in the methodology presented throughout this work if each of the pools is considered as a big static data structure (an array) with a fixed size—however, their approach would prevent splitting a pool over several memory resources because a data structure is viewed as an atomic entity. Nevertheless, that work presents important notions to maximize the chances of parallel accesses when mapping static data structures into several memory modules.

As data objects cannot be usually split during mapping, greedy algorithms based on ordering by frequency of accesses per byte (FPB) are not optimal and most of the works oriented to produce a static data placement resort to more complex methods such as integer linear programming (ILP). This issue is particularly relevant because the amount of individual data objects can increase significantly with the use of dynamic memory (each object-creation location in the code can be executed many times); furthermore, their exact numbers and size are usually not known until run-time. Indeed, the concept of data objects created at run-time is itself not susceptible to such precomputed solutions because further placement choices need to be made when new objects are created, maybe rendering prior decisions undesirable. Either

---

[1]Energy consumption may increase in those cases because the cost of writing and reading input data from the scratchpad, and then writing and reading results before posting them to the DRAM, is added to the cost of accessing the data straight from the DRAM. As a result, independently of whether they are performed by the processor or the DMA, a net overhead of two writes and two reads to the scratchpad is created without any reutilization payback.

prior placement decisions must be lived with, or a data migration mechanism to undo them would be needed.

In contrast with those works, here we propose optimizations for heap data (allocated at run-time) that altogether avoid movements between memory elements that would be difficult to harness due to the lower locality of dynamic data structures. Nevertheless, those approaches are compatible with this work and may complement it for the overall optimization of access to dynamic and static data.

2.3 Dynamic Memory Management

Wide research effort has also been performed on the allocation techniques for dynamic data themselves to construct efficient dynamic memory managers. Several comprehensive surveys have been presented along the years, such as the ones from Margolin et al. [39] and Bozman et al. [12]—which were rather an experimental analysis of many of the then-known memory management techniques with their proposals for improvement in the context of IBM servers—and Wilson et al. [57]—who made a deep analysis of the problem distinguishing between policies and the methods that implement them. Johnstone and Wilson went further in their analysis of fragmentation in general purpose DMMs [28]. They argued that its importance is smaller than previously believed and possibly motivated in big part by the use in early works of statistical distributions that do not reflect the hidden pattern behaviors of real-life applications; thus, they advocated for the use of real-program traces and for the focus of future research to be shifted towards performance. Indeed, the focus of most research has moved from fragmentation and performance to performance and energy consumption; and, since energy consumption depends roughly on the number of instructions and memory accesses executed by the processor, we can say that performance became soon the main goal of most research in the area.

Many works in the field have focused on the design of general-purpose DMMs that could be used under a wide variety of circumstances. In that line, Weinstock and Wulf [56] presented QuickFit, where they proposed to split the heap in two areas, one managed with a set of lists of free blocks and the other with a more general algorithm. Their insight was that most applications (according to their particular experience) allocate blocks from a small set of different sizes, that those sizes are usually small (i.e., they represent small data records) and that applications tend to allocate repeatedly blocks of the same size. Further experiments confirmed that many applications tend to allocate objects of a small number of different sizes, but that those sizes may change significantly between applications. The DMM design presented by Kingsley[2] generalizes that idea and proposes a fast allocation scheme, where a set of lists is created each corresponding to exponentially increasing block

---

[2]See the survey of Wilson et al. [57] for a description of that dynamic memory manager, which was designed for the BSD 4.2 Unix version.

sizes. The design presented by Lea [33] has been the reference design in many Linux distributions for many years as it presents good compromises for different types of applications and their dynamic memory requirements.

Other works have proposed the construction of DMMs specifically tailored to the characteristics of each application. Their most relevant feature is that, contrary to the habit of hand-optimizing the DMMs, they explored automated methods to generate the DMMs after a careful profiling and characterization of the applications. For example, Grunwald and Zorn [27] proposed CustoMalloc, which automatically generates quick allocation lists for the most demanded block sizes in each application, in contrast with the predefined set of lists created by QuickFit. As in that work, a more general allocator takes care of the rest of (less common) allocation sizes. Interestingly, Grunwald and Zorn also reported in their work the presence of some sense of temporal locality for allocations: Applications tend to cluster allocations of the same size.

Vmalloc, presented by Vo [55], proposed an extended interface that applications can use to tune the policies and methods employed by the DMM, providing also a mechanism to generate several regions, each with its own configuration. It also provided auxiliary interfaces for debugging and profiling of programs. Some of his proposals are still offered by systems such as the C run-time library of the Microsoft Visual Studio (debugging and profiling [41]) and the Heap API of Microsoft Windows (separate, application-selectable regions with different method choices).

The set of works presented by Atienza et al. [4–6] and Mamagkakis et al. [36] formalize in a set of decision trees the design space for the construction of DMMs and introduce optimizations specific for the realm of embedded systems that improve on the areas of energy consumption, memory footprint, and performance (cost of the allocation operations). They also defined an ordering between the decision trees to prioritize each of those optimization goals. Their work is complementary to the work we present here as they deal with the design of efficient DMMs and their internal organization, not with the placement problem. Indeed, we rely on it for the step of pool formation in our methodology.

Some effort has also been devoted to partial or full hardware implementations of dynamic memory management. Interesting works in this area are the ones presented by Li et al. [34] and Anagnostopoulos et al. [3].

A common characteristic of all those works is that they do not consider the actual mapping of the pools into physical memory resources. Despite the fact that DDTs become increasingly important as applications in embedded systems grow more complex and driven by external events [37], the heap is frequently left out of the placement optimizations and mapped into the main DRAM. That was the main motivation behind this work.

An interesting work that takes into consideration the characteristics of the memory module where the heap is placed was presented by McIlroy et al. [40], who developed a specialized allocator for scratchpad memories that has its internal data structures highly optimized using bitmaps to reduce their overhead. Their work provided mechanisms to efficiently manage a heap known to reside in a scratchpad memory; it is hence complementary to the proposal of this text, which proposes an efficient placement of dynamic data into the system memories. In other words, here we study the mapping of DDTs into heaps, and of heaps into physical memory resources, whereas their work deals with the internal management of those heaps that have been placed in a scratchpad memory.

Finally, in a different but not less interesting context, Berger et al. [11] introduced a memory allocator for multithreaded applications running on server-class multiprocessor systems. Their allocator makes a careful separation of memory blocks into per-processor heaps and one global heap. However, their goal is not to take advantage of the memory organization, but to avoid the problems of false sharing (of cache lines) and incremented memory consumption under consumer-producer patterns in multiprocessor systems. Although not directly related to the main goal of this work, it is worth mentioning because most middle or high-end embedded devices feature nowadays multiple processors. Exploring the integration of their ideas with techniques for data placement in the context of multiprocessor embedded systems might be an interesting research topic in the immediate future.

2.4   Dynamic Data Types Optimization

Object-oriented languages offer built-in support for DDTs, typically through interfaces for vectors of variable number of elements, lists, queues, trees, maps (associative containers), etc. A software programmer can choose to use the DDT that has the most appropriate operations and data organization, but this is usually done without considering the underlying memory organization. However, changing how the different DDTs are used, considering the data access patterns of the applications and the characteristics of the cache memories, can produce considerable efficiency improvements. Therefore, many authors looked for ways to improve the exploitation of cache resources when using DDTs. One of the resulting works was presented by Chilimbi et al. [15], who applied data optimization techniques such as structure splitting and field reordering to DDTs. A good overview of available transformations was introduced by Daylight et al. [20] and a multi-objective optimization method based on evolutionary computation to optimize complex DDT implementations by Baloukas et al. [9]. Although their authors take the perspective of a programmer implementing the DDTs, it should be fairly easy to apply those techniques to the standard DDTs provided by the language.

A very interesting proposal to improve cache performance was described by Lattner and Adve [32]: A compiler-based approach is taken to build the "points-to" graph of the application DDTs and segregate every single instance of each DDT into a separate pool. However, this approach produces a worst-case assignment of pools to DDTs as the free space in a pool cannot be used to create instances of DDTs from other pools. Most importantly, this work was developed to improve the hit ratio of cache memories (e.g., it enables such clever optimizations as compressing 64-bit pointers into 32-bit integer indexes from the pool base address), but it is not specifically suited for embedded systems with heterogeneous memory organizations. Nevertheless, the possibility of combining their analysis techniques with our placement principles in future works is exciting.

2.5   Dynamic Data Placement

An early approximation to the problem of placement for dynamically allocated data was presented by Avissar et al. [7], but they resorted to a worst-case solution considering each allocation place in the source code as the declaration of a static variable and assigning an upper bound on the amount of memory that can be used by each of them. Moreover, they considered each of these pseudo-static variables as independent entities, not taking into consideration the possibility of managing them in a single pool, and adding considerable complexity to their integer linear optimizer; in that sense, their work lacked a full approach to the concept of DM management. Nonetheless, that work constitutes one of the first approaches to the placement challenge.

Further hints on how to map DDTs into a scratchpad memory were offered by Poletti et al. [45]. In that work, the authors arrived to a satisfactory placement solution that reduces energy consumption and improves performance (albeit the latter only in multiprocessor systems) for a simple case; however, significant manual effort was still required from the designer. The methodology that we present here can produce solutions of similar quality, but in an automated way and with a more global approach that considers all the elements in the memory subsystem (versus a single scratchpad memory).

The work presented by Mamagkakis et al. [36] lays in the boundary between pure DM management, which deals with the problem of efficiently finding free blocks of memory for new objects, and the problem of placing dynamic objects into the right memory modules to reduce the cost of accessing them. There, the authors proposed a method to build DMMs that can be configured to use a specific address range, but leave open the way in which such address range is determined. However, for demonstration purposes, they manually found an object size that received a big fraction of the application data accesses, created a pool specifically for it, and then mapped that pool into a separate scratchpad memory, obtaining important energy and time savings. The work presented in this text is complementary because it tackles with the problem of finding an appropriate placement of data into memory resources, while leaving open the actual mechanism used to allocate blocks inside the pools. Additionally, it classifies the allocations according to not only the size of the memory blocks, but also the high-level data type of the objects. This enables a complete analysis of the DDTs that is impossible—or, at least, very difficult—if different DDTs with the same size are not differentiated.

More recently, efforts to create an algorithm to map dynamic, linked, data structures to a scratchpad memory have been presented by Domínguez et al. [21] and Udayakumaran et al. [52], who propose to place in the scratchpad some portions of the heap, called "bins." The bins are moved from the main memory to the scratchpad when the data they hold are known to be accessed at the next execution point of the application. For each DDT, a bin is created and only its first instances will reside in it (and so in the scratchpad), whereas the rest will be kept in a different pool permanently mapped into the main memory. That method requires a careful analysis of the application and instrumentation of the final code to execute the data movements properly. One drawback of this approach is that in order to offset the cost of the data movements, the application must spend enough time in the region of code that benefits from the new data distribution, reusing the data in the bins. Compared to this approach, the method presented here avoids data migration between elements of the memory subsystem and considers the footprint of all the instances of the DDTs, not only the first ones.

The work that we describe here has three important differences in relation with previous techniques. First, it tries to solve the problem of mapping all the DDTs, not only into one scratchpad memory, but into all the different elements of a heterogeneous memory subsystem. Second, it employs an exclusive memory organization model [29], which has some advantages under certain circumstances [51, 59]. As applied here, this model avoids duplication of data across different levels of the memory subsystem: Each level holds distinct data and no migrations are performed. Avoiding data movements reduces the energy and cycles overhead at the possible cost of using less efficiently the memory resources during specific phases of the application execution. However, due to the additional grouping step, which is based on the analysis of the memory footprint evolution and access characteristics of each DDT, and the fact that no resources are wasted in duplicated data—effectively increasing the usable size of the platform memories—we argue that this methodology can overcome the possible inefficiencies in many situations. As a final consideration, this method is also applicable to multithreaded applications where DDTs are accessed from different threads.

In a different order of things, several algorithms to perform an efficient mapping of dynamic applications (i.e., triggered by unpredictable external events) into DRAM modules were presented by Marchal et al. [38]. However, they dealt with other aspects of the applications' unpredictability, not with the DDTs created in the heap. Interestingly, they introduced the concept of selfishness to reduce the number of row misses in the banks of each DRAM module. That idea could be easily integrated into this work to control the mapping of DDTs on the different banks of DRAM modules.

2.6 Computational Complexity

The field of computational complexity is broad, but little is required to understand the work presented here. A good introduction can be obtained from the classic textbook of Cormen et al. [18, Chap. 34]. Our work is closely related to the family of problems that contains the knapsack and general assignment problems. A very good introduction to those problems was presented by Pisinger in his PhD work [43] and several related works where he studied specific instances such as the multiple knapsack problem [44].

Chekuri and Khanna [14] present a polynomial-time approximation scheme (PTAS) for the multiple knapsack problem. They also propose that this is the most complex special case of GAP that is not APX-hard—i.e., that is not "hard" even to approximate. In the context of our own work, this means that for placement, which is more complex, no PTAS is likely to be found. Figure 1 in their work shows a brief schematic of the approximability properties of different variations of the knapsack/GAP problems.

Shmoys and Tardos [47] present the minimization version of GAP (Min GAP) and propose several approximation algorithms for different variations of that problem, which is relevant to our work because placement represents an effort of minimization and the maximization and minimization versions of the problem are not exactly equivalent—at least as far as we know. Many works deal with the fact that the general versions of GAP are even hard to approximate. In that direction, Cohen et al. [16] explore approximation algorithms for GAP based on approximation algorithms for knapsack problems, while Fleischer et al. [23] present an improvement on previous approximation boundaries for GAP.

Finally, virtual machine colocation, which has strong resemblances with the data placement problem, was recently studied by Sindelar et al. [48]. The particularity of this problem is that different virtual machines that run similar operating systems and applications usually have many pages of memory that are exactly identical (e.g., for code sections of the operating system). Placing them in the same physical server allows the hypervisor to reduce their total combined footprint.

References

1. Absar, M., Poletti, F., Marchal, P., Catthoor, F., Benini, L.: Fast and power-efficient dynamic data-layout with DMA-capable memories. In: Proceedings of the International Workshop on Power-Aware Real-Time Computing (PACS) (2004)
2. Ahmed, N., Mateev, N., Pingali, K.: Tiling imperfectly-nested loop nests. In: Proceedings of the IEEE/ACM International Conference for High Performance Computing, Networking, Storage, and Analysis (SC). IEEE Computer Society Press, Dallas (2000). http://dl.acm.org/citation.cfm?id=370049.370401
3. Anagnostopoulos, I., Xydis, S., Bartzas, A., Lu, Z., Soudris, D., Jantsch, A.: Custom microcoded dynamic memory management for distributed on-chip memory organizations. IEEE Embed. Syst. Lett. 3(2), 66–69 (2011). https://doi.org/10.1109/LES.2011.2146228
4. Atienza, D.: Metodología multinivel de refinamiento del subsistema de memoria dinámica para los sistemas empotrados multimedia de altas prestaciones. Ph.D. thesis, Universidad Complutense de Madrid, Departamento de Arquitectura de Computadores y Automática (2005)
5. Atienza, D., Mamagkakis, S., Catthoor, F., Mendías, J.M., Soudris, D.: Dynamic memory management design methodology for reduced memory footprint in multimedia and wireless network applications. In: Proceedings of Design, Automation and Test in Europe (DATE), vol. 1, pp. 532–537 (2004). https://doi.org/10.1109/DATE.2004.1268900

6. Atienza, D., Mendías, J.M., Mamagkakis, S., Soudris, D., Catthoor, F.: Systematic dynamic memory management design methodology for reduced memory footprint. ACM Trans. Des. Autom. Electron. Syst. **11**(2), 465–489 (2006) https://doi.org/10.1145/1142155.1142165

7. Avissar, O., Barua, R., Stewart, D.: Heterogeneous memory management for embedded systems. In: Proceedings of the International Conference on Compilers, Architectures and Synthesis for Embedded Systems (CASES), pp. 34–43. ACM Press, Atlanta (2001). https://doi.org/10.1145/502217.502223

8. Banakar, R., Steinke, S., Lee, B.S., Balakrishnan, M., Marwedel, P.: Scratchpad memory: a design alternative for cache on-chip memory in embedded systems. In: Proceedings of the International Symposium on Hardware/Software Codesign (CODES), pp. 73–78. ACM Press, Estes Park (2002). https://doi.org/10.1145/774789.774805

9. Baloukas, C., Risco-Martín, J.L., Atienza, D., Poucet, C., Papadopoulos, L., Mamagkakis, S., Soudris, D., Hidalgo, J.I., Catthoor, F., Lanchares, J.: Optimization methodology of dynamic data structures based on genetic algorithms for multimedia embedded systems. J. Syst. Softw. **82**(4), 590–602 (2009). https://doi.org/10.1016/j.jss.2008.08.032. Special Issue: Selected papers from the 2008 IEEE Conference on Software Engineering Education and Training (CSEET'08)

10. Benini, L., Micheli, G.: System-level power optimization: techniques and tools. ACM Trans. Design Autom. Electron. Syst. **5**(2), 115–192 (2000). https://doi.org/10.1145/335043.335044

11. Berger, E.D., McKinley, K.S., Blumofe, R.D., Wilson, P.R.: Hoard: a scalable memory allocator for multithreaded applications. ACM SIGPLAN Notices **35**(11), 117–128 (2000). https://doi.org/10.1145/356989.357000

12. Bozman, G., Buco, W., Daly, T.P., Tetzlaff, W.H.: Analysis of free-storage algorithms— revisited. IBM Syst. J. **23**(1), 44–64 (1984). https://doi.org/10.1147/sj.231.0044

13. Catthoor, F., Wuytack, S., Greef, E., Balasa, F., Nachtergaele, L., Vandecappelle, A.: Custom Memory Management Methodology: Exploration of Memory Organisation for Embedded Multimedia System Design. Kluwer Academic Publishers, Boston (1998)

14. Chekuri, C., Khanna, S.: A PTAS for the multiple knapsack problem. In: Proceedings of the Annual ACM-SIAM Symposium on Discrete Algorithms (SODA), pp. 213–222 (2000). http://dl.acm.org/citation.cfm?id=338219.338254

15. Chilimbi, T.M., Davidson, B., Larus, J.R.: Cache-conscious structure definition. In: Proceedings of the ACM SIGPLAN Conference on Programming Language Design and Implementation (PLDI), pp. 13–24. ACM Press, Atlanta (1999). https://doi.org/10.1145/301618.301635

16. Cohen, R., Katzir, L., Raz, D.: An efficient approximation for the Generalized Assignment Problem. Inf. Process. Lett. **100**(4), 162–166 (2006). https://doi.org/10.1016/j.ipl.2006.06.003

17. Coleman, S., McKinley, K.S.: Tile size selection using cache organization and data layout. In: Proceedings of the ACM SIGPLAN Conference on Programming Language Design and Implementation (PLDI), pp. 279–290. ACM Press, La Jolla (1995). https://doi.org/10.1145/207110.207162

18. Cormen, T.H., Leiserson, C.E., Rivest, R.L., Stein, C.: Introduction to Algorithms, 2nd edn. MIT Press and McGraw-Hill Book Company (2001)

19. Dasygenis, M., Brockmeyer, E., Durinck, B., Catthoor, F., Soudris, D., Thanailakis, A.: A combined DMA and application-specific prefetching approach for tackling the memory latency bottleneck. IEEE Trans. Very Large Scale Integr. Syst. **14**(3), 279–291 (2006). https://doi.org/10.1109/TVLSI.2006.871759

20. Daylight, E., Atienza, D., Vandecappelle, A., Catthoor, F., Mendías, J.M.: Memory-access-aware data structure transformations for embedded software with dynamic data accesses. IEEE Trans. Very Large Scale Integr. Syst. **12**(3), 269–280 (2004). https://doi.org/10.1109/TVLSI.2004.824303

21. Domínguez, A., Udayakumaran, S., Barua, R.: Heap data allocation to scratch-pad memory in embedded systems. J. Embedded Comput. **1**(4), 521–540 (2005)

22. Duch, L., Basu, S., Braojos, R., Ansaloni, G., Pozzi, L., Atienza, D.: HEAL-WEAR: an ultra-low power heterogeneous system for bio-signal analysis. IEEE Trans. Circuits Syst. I Regul. Pap. **64**(9), 2448–2461 (2017)

23. Fleischer, L., Goemans, M.X., Mirrokni, V.S., Sviridenko, M.: Tight approximation algorithms for maximum general assignment problems. In: Proceedings of the Annual ACM-SIAM Symposium on Discrete Algorithms (SODA), pp. 611–620 (2006). http://dl.acm.org/citation. cfm?id=1109557.1109624

24. Geelen, B., Brockmeyer, E., Durinck, B., Lafruit, G., Lauwereins, R.: Alleviating memory bottlenecks by software-controlled data transfers in a data-parallel wavelet transform on a multicore DSP. In: Proceedings of the IEEE BENELUX/DSP Valley Signal Processing Symposium (SPS-DARTS), pp. 143–146 (2005)

25. González-Alberquilla, R., Castro, F., Piñuel, L., Tirado, F.: Stack filter: reducing L1 data cache power consumption. J. Syst. Archit. **56**(12), 685–695 (2010). https://doi.org/10.1016/j.sysarc. 2010.10.002

26. GreenWaves Technologies: GAP8 Auto-tiler manual. https://greenwaves-technologies.com/ manuals/BUILD/AUTOTILER/html/index.html. Accessed May 2019

27. Grunwald, D., Zorn, B.: CustoMalloc: efficient synthesized memory allocators. Softw. Pract. Exp. **23**, 851–869 (1993). https://doi.org/10.1002/spe.4380230804

28. Johnstone, M.S., Wilson, P.R.: The memory fragmentation problem: solved? In: Proceedings of the International Symposium on Memory Management (ISMM), pp. 26–36. ACM Press, Vancouver (1998). https://doi.org/10.1145/286860.286864

29. Jouppi, N.P., Wilton, S.J.E.: Tradeoffs in two-level on-chip caching. In: Proceedings of the International Symposium on Computer Architecture (ISCA), pp. 34–45. IEEE Computer Society Press, Chicago (1994). https://doi.org/10.1145/191995.192015

30. Kandemir, M., Ramanujam, J., Irwin, J., Vijaykrishnan, N., Kadayif, I., Parikh, A.: Dynamic management of scratch-pad memory space. In: Proceedings of the Design Automation Conference (DAC), pp. 690–695 (2001). https://doi.org/10.1145/378239.379049

31. Kandemir, M., Kadayif, I., Choudhary, A., Ramanujam, J., Kolcu, I.: Compiler-directed scratchpad memory optimization for embedded multiprocessors. IEEE Trans. Very Large Scale Integr. Syst. **12**, 281–287 (2004)

32. Lattner, C., Adve, V.: Automatic pool allocation: improving performance by controlling data structure layout in the heap. In: Proceedings of the ACM SIGPLAN Conference on Programming Language Design and Implementation (PLDI), pp. 129–142. ACM Press, Chicago (2005). https://doi.org/10.1145/1065010.1065027

33. Lea, D.: A memory allocator (1996). http://g.oswego.edu/dl/html/malloc.html

34. Li, W., Mohanty, S., Kavi, K.: A page-based hybrid (software-hardware) dynamic memory allocator. IEEE Comput. Archit. Lett. **5**(2), 13–13 (2006)

35. Lim, A.W., Liao, S.W., Lam, M.S.: Blocking and array contraction across arbitrarily nested loops using affine partitioning. In: Proceedings of the ACM SIGPLAN Symposium on Principles and Practices of Parallel Programming (PPoPP), pp. 103–112. ACM Press, Snowbird, Utah (2001). https://doi.org/10.1145/379539.379586

36. Mamagkakis, S., Atienza, D., Poucet, C., Catthoor, F., Soudris, D.: Energy-efficient dynamic memory allocators at the middleware level of embedded systems. In: Proceedings of the ACM and IEEE International Conference on Embedded Software (EMSOFT), vol. 1, pp. 215–222. ACM Press, Seoul (2006). https://doi.org/10.1145/1176887.1176919

37. Man, H.: Connecting e-dreams to deep-submicron realities. In: Proceedings of International Workshop on Power And Timing Modeling, Optimization and Simulation (PATMOS). Springer, Berlin (2004). https://doi.org/10.1007/b100662

38. Marchal, P., Catthoor, F., Bruni, D., Benini, L., Gómez, J.I., Piñuel, L.: Integrated task scheduling and data assignment for SDRAMs in dynamic applications. IEEE Design Test Comput. **21**(5), 378–387 (2004). https://doi.org/10.1109/MDT.2004.66

39. Margolin, B.H., Parmelee, R.P., Schatzoff, M.: Analysis of free-storage algorithms. IBM Syst. J. **10**(4), 283–304 (1971). https://doi.org/10.1147/sj.104.0283

40. McIlroy, R., Dickman, P., Sventek, J.: Efficient dynamic heap allocation of scratch-pad memory. In: Proceedings of the International Symposium on Memory Management (ISMM), pp. 31–40. ACM Press, Tucson (2008). https://doi.org/10.1145/1375634.1375640

41. Microsoft Corporation: CRT debug heap details (Last fetched on November 2018). https://msdn.microsoft.com/en-us/library/974tc9t1.aspx
42. Panda, P.R., Dutt, N.D., Nicolau, A.: On-chip vs. off-chip memory: the data partitioning problem in embedded processor-based systems. ACM Trans. Design Autom. Electron. Syst. **5**(3), 682–704 (2000). https://doi.org/10.1145/348019.348570
43. Pisinger, D.: Algorithms for knapsack problems. Ph.D. thesis, University of Copenhagen (1995)
44. Pisinger, D.: An exact algorithm for large multiple knapsack problems. Eur. J. Oper. Res. **114**(3), 528–541 (1999). https://doi.org/10.1016/S0377-2217(98)00120-9
45. Poletti, F., Marchal, P., Atienza, D., Benini, L., Catthoor, F., Mendías, J.M.: An integrated hardware/software approach for run-time scratchpad management. In: Proceedings of the Design Automation Conference (DAC) (2004)
46. Rossi, D., Loi, I., Conti, F., Tagliavini, G., Pullini, A., Marongiu, A.: Energy efficient parallel computing on the PULP platform with support for OpenMP. In: IEEE Convention of Electrical Electronics Engineers in Israel (IEEEI), pp. 1–5 (2014). https://doi.org/10.1109/EEEI.2014.7005803
47. Shmoys, D.B., Tardos, E.: An approximation algorithm for the generalized assignment problem. Math. Program. **62**, 461–474 (1993)
48. Sindelar, M., Sitaraman, R., Shenoy, P.: Sharing-aware algorithms for virtual machine colocation. In: Proceedings of the ACM Symposium on Parallelism in Algorithms and Architectures (SPAA), pp. 367–378 (2011)
49. Soto, M., Rossi, A., Sevaux, M.: A mathematical model and a metaheuristic approach for a memory allocation problem. J. Heuristics **18**(1), 149–167 (2012). https://doi.org/10.1007/s10732-011-9165-3
50. Steinke, S., Wehmeyer, L., Lee, B., Marwedel, P.: Assigning program and data objects to scratchpad for energy reduction. In: Proceedings of Design, Automation and Test in Europe (DATE), p. 409 (2002)
51. Subha, S.: An exclusive cache model. In: International Conference on Information Technology: New Generations (ITNG), pp. 1715–1716. IEEE Computer Society Press, Las Vegas (2009). https://doi.org/10.1109/ITNG.2009.89
52. Udayakumaran, S., Domínguez, A., Barua, R.: Dynamic allocation for scratch-pad memory using compile-time decisions. ACM Trans. Embedded Comput. 5(2), 472–511 (2006). https://doi.org/10.1145/1151074.1151085
53. Verma, M., Steinke, S., Marwedel, P.: Data partitioning for maximal scratchpad usage. In: Proceedings of the Asia and South Pacific Design Automation Conference (ASP-DAC), pp. 77–83 (2003). https://doi.org/10.1145/1119772.1119788
54. Verma, M ., Wehmeyer, L., Marwedel, P.: Cache-aware scratchpad allocation algorithm. In: Proceedings of Design, Automation and Test in Europe (DATE) (2004)
55. Vo, K.P.: Vmalloc: a general and efficient memory allocator. Softw. Pract. Exp. **26**(3), 357–374 (1996). https://doi.org/10.1002/(SICI)1097-024X(199603)26:3<357::AID-SPE15>3.0.CO;2-#
56. Weinstock, C.B., Wulf, W.A.: Quick fit: an efficient algorithm for heap storage allocation. ACM SIGPLAN Notices **23**(10), 141–148 (1988). https://doi.org/10.1145/51607.51619
57. Wilson, P.R., Johnstone, M.S., Neely, M., Boles, D.: Dynamic storage allocation: a survey and critical review. In: Proceedings of the International Workshop on Memory Management (IWMM), pp. 1–116. Springer, Berlin (1995). http://dl.acm.org/citation.cfm?id=645647.664690
58. Wuytack, S., Diguet, J.P., Catthoor, F., Man, H.: Formalized methodology for data reuse exploration for low-power hierarchical memory mappings. IEEE Trans. Very Large Scale Integr. Syst. **6**(4), 529–537 (1998). https://doi.org/10.1109/92.736124
59. Zheng, Y., Davis, B.T., Jordan, M.: Performance evaluation of exclusive cache hierarchies. In: Proceedings of the IEEE International Symposium on Performance Analysis of Systems and Software (ISPASS), pp. 89–96. IEEE Computer Society Press, Washington (2004). https://doi.org/10.1109/ISPASS.2004.1291359

Chapter 3
A Gentle Introduction to Dynamic Memory, Linked Data Structures and Their Impact on Access Locality

3.1 Use Cases for Dynamic Memory

Dynamic memory (DM) has two main components: A range of reserved memory addresses and the algorithms needed to manage it. The algorithms perform their bookkeeping building data structures that are usually embedded in the address range itself. As in the rest of this work, we use the term *heap* to denote the address range, *dynamic memory manager* (DMM) for the algorithms, and *pool* for the algorithms, their internal control data structures and the address range as a whole. With these definitions, let us consider now three common usage patterns of dynamic memory that applications can employ to organize their internal data objects.

3.1.1 Allocation of Vectors Whose Size Is Known Only at Run-Time

This is the most basic case and, apart from the allocation and deallocation calls themselves, the rest of the process is exactly the same as accessing a vector through a pointer in C or C++. In Java, this is the standard implementation for the allocation of arrays, either of primitive types or objects. The following code snippet shows how a vector can be created and destroyed dynamically, and a possible layout in memory. The pointer to the starting address of the vector itself is usually located either in the stack or in the global data segment; in this example, it resides in the stack:

© Springer Nature Switzerland AG 2020
M. Peón Quirós et al., *Heterogeneous Memory Organizations in Embedded Systems*, https://doi.org/10.1007/978-3-030-37432-7_3

```
double * theVector = NULL;
int vectorSize = DetermineInputDataSize();

theVector = new double[vectorSize];

/*... Vector elements are used here ...*/

delete[] theVector;
```

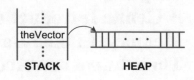

Vectors allocated in this way are cache-friendly almost as if they were statically allocated vectors, particularly if there are lots of sequential accesses to the vector once it is allocated. The main difference appears if the application creates and destroys multiple instances of small vectors. If a static vector existed, then all the associated memory positions could remain loaded in the corresponding cache lines and directly accessed every time the application used the vector. However, every new instance of a dynamically allocated vector can be created in a different memory position (e.g., with a dynamic memory manager that employs a FILO policy for block reuse), even if only one is alive at a time, which increases the number of potential cache misses. This increases also the chances of an address collision with other data objects that forces evictions of cache lines, increasing the traffic between elements in the memory subsystem.

Several previous works focused on adapting techniques already developed for the management of static data (i.e., statically allocated during the design phase either in the global data segment or in the stack) to dynamically allocated vectors such as the ones explained here. For instance, [6] explored the possibility of allocating arrays (vectors) at run-time to exploit the characteristics of multi-banked DRAMs.

3.1.2   Creation of a Collection of Data Objects with an Unknown Cardinality

The previous point shows the case in which the size of a vector is unknown; however, the number of vectors required may also be unknown, for example, if they are created inside a loop with a data-dependent number of iterations. In those cases, instead of creating a vector of objects with a worst-case number of entries, the programmer can create a vector of pointers that acts as an index. The objects will be created and linked as needed. This technique, which can be generalized to any type of structures or objects, may reduce the memory footprint of the application significantly because it allocates space only for the objects that are currently needed, plus the size of the vector of pointers. The following code snippet illustrates the concept with a vector that holds pointers to other dynamic vectors of varying sizes that are in turn created and destroyed dynamically:

```
// vectorSizes[ii] contains the size of the
// vector at theVectors[ii]
int numVectors = 0;
int * vectorSizes = NULL;
double ** theVectors = NULL;

numVectors = CalcNumberOfInputVectors();
vectorSizes = new int[numVectors];
theVectors = new double * [numVectors];

for (int ii = 0; ii < numVectors; ++ ii)
{
      vectorSizes[ii] = GetVectorSize(ii);
      theVectors[ii] =
              new double[vectorSizes[ii]];
}

/*... Do something useful ...*/

/* Delete all the vectors */
for (int ii = 0; ii < numVectors; ++ ii)
      delete[] theVectors[ii];
delete[] theVectors;
delete[] vectorSizes;
```

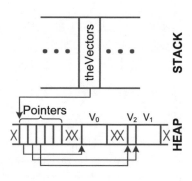

In this example, the number of elements in each of the dynamic vectors is known when the vectors are created: The vectors cannot grow or decrease once they are allocated until they are destroyed. However, the application can destroy an individual vector and create a new one with a different size, updating the pointer in the index.

3.1.3 Construction of Dynamic Data Types (DDTs)

DDTs are usually complex structures of nodes "linked" through pointers, such as lists, queues, trees, tries, etc., and whose components are not necessarily in contiguous memory addresses. Many languages such as C++, Java, Python, or various domain-specific languages (DSLs) offer a standard library of ready-to-use DDTs that usually includes iterator-based or associative containers and provides a smooth method to group data objects, hence providing access to dynamic memory at a high-level of abstraction. A relevant property of DDTs is that the access time to the elements is potentially variable. For example, in a linked list the application has to traverse the $n - 1$ first elements before getting access to the nth one.

This case is the most flexible because it can deal both with unknown size and with unknown cardinality in the application data. The DDT itself is constructed as new objects are created; therefore, no worst-case provisions are needed. The following fragment of code shows a simplified example for the construction of a linked list. Each of the nodes may be physically allocated at any possible position in the heap:

```
struct TNodo {
      TNodo * next;
      int data;
};

TNodo * first = NULL;
TNodo * it = NULL, * aux = NULL;

first = new TNodo;
first->next = NULL;
first->data = 0;

/* Add some elements */
it = first;
while (MoreElements()) {
      it->next = new TNodo;
      it = it->next;
      it->next = NULL;
      it->data = GetDataElement();
}

/*... Do something useful ...*/

/* Delete elements */
it = first;
while (it != NULL) {
      aux = it;
      it = it->next;
      delete aux;
}
```

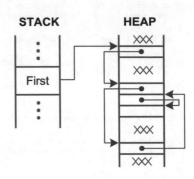

This model can be composed as needed, for instance, building a vector of pointers to a previously unknown number of lists each with an indeterminate number of elements, or building lists of lists if the number of lists has to be adapted dynamically. The possible combinations are endless. Figure 3.1 shows examples of different DDTs and their organizations. The nodes of the DDTs may contain the data values themselves, or they may store a pointer to another dynamic object that represents the actual data. The second possibility requires additional memory to store the pointer to the object (plus the memory overhead introduced by the DMM to allocate it), but it offers advantages such as the possibility of creating variable-sized objects, creating empty nodes (e.g., through the use of NULL pointers), and separating the placement of the DDT nodes from the placement of the contained objects [3, 4]. This last transformation can improve data structure traversals because the nodes, which hold the pointers to the next ones and may participate in traversals, can be tightly packed into an efficient memory, while the data elements, which may be much bigger and less frequently accessed, are stored in a different one.

As a final remark, although the concept of DDT is applied to the whole dynamic structure, the allocation and placement are executed for every instance of each node in the DDT.

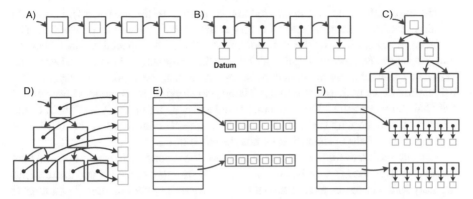

Fig. 3.1 Several examples of DDTs: (**a**) Linked list with data in the nodes. (**b**) Linked list with pointers to external objects in the nodes. (**c**) Tree with data in the nodes. (**d**) Tree with pointers to external objects in the nodes. (**e**) Open hash table with data in the nodes of each entry's list. (**f**) Open hash table with pointers to external data in the nodes of each entry's list

3.2 Impact of Linked Data Structures on Locality

Let us explore now three examples of increasing complexity that illustrate different situations that arise during the utilization of DDTs and their possible impact on cache memories. The first one uses a linked list to show the effect of element removal and the interactions with the state of the dynamic memory manager (DMM). The second example uses an AVL tree to explain that just the internal organization of the DDT can alter the locality of elements, even without deletions, and how different traversals affect also spatial locality. Finally, the third example justifies that the behavior observed with the AVL trees appears also when working with random data and explains the relation between tree nodes and cache lines.

3.2.1 Example: A Simple Case with the List DDT

Let us consider an application that uses a linked list. We can assume for now that the DMM keeps a list of free blocks. Every time the application needs to insert a new node in the list, it issues an allocation request to the DMM. If the request can be served with any of the previously existing free blocks, then that block is assigned. Otherwise, the DMM gets a new block of memory from the system resources (e.g., using a call to mmap or sbrk). In this simple scenario, the first nodes allocated by the application may get consecutive memory addresses (they are allocated consecutively from a single block of system resources):

The first node of the list is created and its pointer to the next element ("Next") is initialized to NULL to signal that this is also the last element of the list. The application keeps track of the first node in the list with a special pointer, "Head," that can be created as any regular global, stack, or dynamically allocated variable. When new elements are appended at the end of the list, the application traverses the list starting at the node pointed to by "Head" and stopping at the node whose "Next" field has the value NULL. The new node is linked by updating the "Next" field of the formerly last node with the address of the new one. The invariant of this DDT says that the "Next" field of the last node has the value NULL.

At a later point, the application does not need the second node any longer and thus destroys it. The node is unlinked from the logical structure of the list and the released memory space becomes available to the DMM for assignment in later allocations. It becomes immediately noticeable that the first and (now) second nodes are no longer situated in consecutive memory addresses:

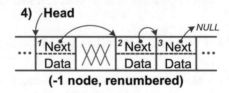

(-1 node, renumbered)

Next, the application receives a new data element and needs to append it at the end of the linked list. However, the DMM has now one suitable block in the list of free blocks. Therefore, it uses that block to satisfy the new memory request:

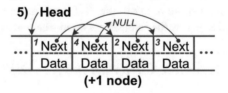

(+1 node)

This simple example shows that although the logical structure of the DDT is preserved, a single element removal alters the spatial locality of the list nodes. After a number of operations, each consecutive node may be in a different cache line. This effect may lead to increased traffic between the cache and main memory and unwanted interactions among nodes of different DDTs. Furthermore, the spreading of logical nodes across memory addresses does not only depend on the operations performed on the DDT; it depends also on the previous state of the DMM and the interactions with operations performed on other DDTs.

The effects of list traversals on the performance of cache memories are an interesting topic. In the worst case, one data node will contain a pointer to the next node and a pointer to the data element corresponding to that node (or a single integer number). Assuming a 32-bit architecture with 64-byte cache lines, every cache line will have 64 bits of useful data for every 512 bits of data storage (or two 32-bit words for every sixteen words of storage). If the number of elements in a list becomes sufficiently large so that every node access requires fetching a new cache line from main memory during every traversal, then an 87.5% of the data are transferred without benefit. A careful programmer or compiler can insert prefetch instructions to hide the time required for the data transfers; if the address of the next node can be calculated quickly and there is enough work to perform on every node, then the delay may be completely hidden. However, energy consumption is a different story: Every data transfer sips a bit of energy, regardless of whether the data is later employed or not. Even worst, as the list grows, every traversal can evict a potentially large fraction of the data previously held by the cache.

3.2.2   A More Complex Example with AVL Trees

Some data structures depend on a correct organization of their internal nodes to guarantee predictable complexity orders. For example, search trees [2] require a certain balance between the left and right branches of every node in order to guarantee a search time in the order of $O(\log_2 n)$. However, the order in which the data elements are inserted and their actual values influence the internal organization of the dynamic data structure itself. Self-balanced structures such as AVL or red-black trees have been designed to palliate this problem. An AVL tree [1] is a binary search tree in which the difference between the weights of the children is at most one for every node. AVL trees keep the weight balance recursively. If any action on the tree changes this condition for a node, then an operation known as a "rotation" [2] is performed to restore the equilibrium. These rotations introduce an additional cost; however, under the right circumstances, it is distributed among all the operations, yielding an effective *amortized cost* of $O(\log_2 n)$. How does that behavior relate to the problem presented in this text? The internal reorganization of the nodes in the AVL tree changes the logical relations between them, but their placement (memory addresses) was already fixed by the DMM at allocation time.

Consider now the case of an AVL tree with the following definition:

Offset	Size	Field declaration
0	4	UINT32 key
4	4	TAVLNode * parent
8	4	TAVLNode * leftChild
12	4	TAVLNode * rightChild
16	4	TData * data
20	1	INT8 balance

Assuming 32-bit pointers and 32-bit padding for the last field, the size of the nodes is 24 B. We can study the construction of the tree as the integer numbers from 1 to 12 are inserted in order: 1, 2, 3, 4, 5, 6, 7, 8, 9, 10, 11, and 12. First, the number "1" is inserted in the empty tree:

The black number (left) represents the value stored in the node. The red number (right, up) represents the balance factor: -1 if the left child has a bigger weight, $+1$ if the right child has a bigger weight, 0 if both children have the same weight. An absolute value larger than 1 means that the node is unbalanced and a rotation must be performed. Finally, the green number (right, bottom) represents the memory address of the node, as assigned by a hypothetical DMM manager.

When number "2" is inserted, it goes to the right child, as is customary in binary search trees for values bigger than the value in the root node:

The new node is balanced because both (null) children have equal weight. However, the root node is unbalanced towards the right child, although still inside the allowed margin. When number "3" is inserted, it goes to the right child of the root node because its value is bigger. Then, as it is also bigger than "2," it goes again to the right child. The root node becomes completely unbalanced (weight $+2$). In this case, a simple rotation towards the left is enough to restore the balance in the tree:

After the previous rotation, the balance factor of all the nodes is restored to 0. When the value "4" is added to the tree, it becomes again the rightmost child of the whole tree. Notice the new weight balances along the tree. The node that contains the value "3" has a right son and no left son, so it has a balance of $+1$. The node "2" has a right son with a depth of 2 and a left son with a depth of 1; therefore, this node has a balance of $+1$. The tree is still globally balanced:

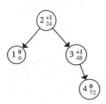

Adding the next value, "5," unbalances the tree again. This time, the node "3" has a right son of depth 2 and a left son of depth 0, so its own balance is +2. A simple rotation is again enough to fix the subtree[1]:

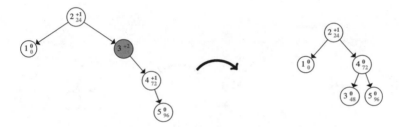

When "6" is added to the tree, the situation gets a bit more complex. The value is again added as a node in the rightmost part of the tree. This time the root of the tree itself becomes unbalanced: Its right child has a maximum depth of three levels, while its left child has a depth of 1, giving a total balance of +2. However, a simple rotation as the ones performed before is not enough as it would leave "4" at the root, but with three children: "2," "3," and "5." A *complex rotation* is needed in this case.

After the rotation, node "4" becomes the new root and node "2" its left child. Additionally, the former left child of "4" becomes now the right child of "2." This is acceptable because binary search trees require only that the values of a node's left children are smaller than its own value and the values of all right children, bigger. In the original tree, node "3" was on the right part of "2," so it was bigger than it. In the new tree, it is also at the right side of "2," preserving the requirement. In respect to "4," which becomes the new root of the tree, both nodes, "2" and "3," are smaller than it, so they can be organized in any way as long as they are both in the left part of the tree. After this new type of rotation, the global tree becomes again balanced:

---

[1] It can be formally proven that the rotations described here restore the global balance of the tree without the need for more rotations in the upper levels of the tree. However, this proof is out of the scope of this work. Further references can be found in the literature, for instance, in [1, 2, 7].

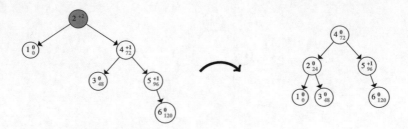

Adding the values from "7" to "11" requires a simple rotation, nothing, a simple rotation, a complex rotation, and a simple rotation, respectively:

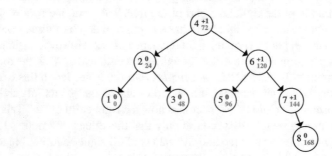

Adding value "7" and rotating.

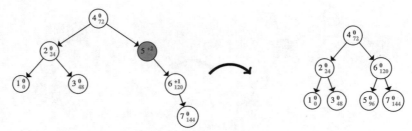

Adding value "8."

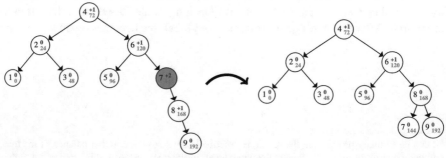

Adding node "9" and rotating.

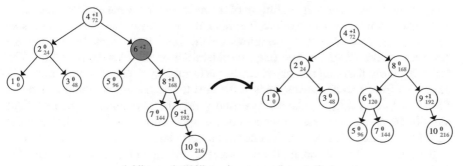

Adding node "10" requires a complex rotation.

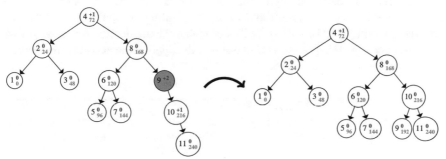

Adding node "11" and rotating.

After the (complex) rotation required to add "12," the final configuration of the AVL tree becomes:

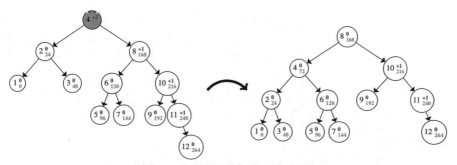

Adding node "12" and final configuration.

Assuming that the DMM assigns consecutive addresses to the nodes as they are created, the final layout of the tree in memory is:

1 0	2 24	3 48	4 72	5 96	6 120	7 144	8 168	9 192	10 216	11 240	12 264

Up to here, we have seen a glimpse of the mismatch between memory addresses and logical links. We can explore it further if we analyze the memory accesses needed to perform different operations on the tree. First, consider a complete "in-order" traversal of the tree (e.g., to obtain the ordered list of elements). The following table lists the memory accesses performed by the application. Each cell, numbered from 1 to 23, corresponds to the visit to one node. The first line in a cell shows the node value, its address on memory, and the step number. The next lines identify the memory accesses executed by the application while visiting that node: "L" if the application reads the pointer to the left child, "R" if it reads the pointer to the right one, "D" if the application accesses the data element at the node, "P" if the pointer to the parent node is used to go up one level in the tree, and "K" if the application reads the key of the node. For each access, the table shows the offset from the start of the node and the corresponding absolute memory address. For example, the line "L: $+8 \rightarrow 176$" means that the application reads the pointer to the left node; as this pointer is at offset $+8$, the application accesses the memory word at position $168 + 8 = 176$.

8_{168} (1)	4_{72} (2)	2_{24} (3)	1_{0} (4)	2_{24} (5)	3_{48} (6)
L: $+8 \rightarrow 176$	L: $+8 \rightarrow 80$	L: $+8 \rightarrow 32$	L: $+8 \rightarrow 8$	D: $+16 \rightarrow 40$	L: $+8 \rightarrow 56$
			D: $+16 \rightarrow 16$	R: $+12 \rightarrow 36$	D: $+16 \rightarrow 64$
			R: $+12 \rightarrow 12$		R: $+12 \rightarrow 60$
			P: $+4 \rightarrow 4$		P: $+4 \rightarrow 52$
2_{24} (7)	4_{72} (8)	6_{120} (9)	5_{96} (10)	6_{120} (11)	7_{144} (12)
P: $+4 \rightarrow 28$	D: $+16 \rightarrow 88$	L: $+8 \rightarrow 128$	L: $+8 \rightarrow 104$	D: $+16 \rightarrow 136$	L: $+8 \rightarrow 152$
	R: $+12 \rightarrow 84$		D: $+16 \rightarrow 112$	R: $+12 \rightarrow 132$	D: $+16 \rightarrow 160$
			R: $+12 \rightarrow 108$		R: $+12 \rightarrow 156$
			P: $+4 \rightarrow 100$		P: $+4 \rightarrow 148$
6_{120} (13)	4_{72} (14)	8_{168} (15)	10_{216} (16)	9_{192} (17)	10_{216} (18)
P: $+4 \rightarrow 124$	P: $+4 \rightarrow 76$	D: $+16 \rightarrow 184$	L: $+8 \rightarrow 224$	L: $+8 \rightarrow 200$	D: $+16 \rightarrow 232$
		R: $+12 \rightarrow 180$		D: $+16 \rightarrow 208$	R: $+12 \rightarrow 228$
				R: $+12 \rightarrow 204$	
				P: $+4 \rightarrow 196$	
11_{240} (19)	12_{264} (20)	11_{240} (21)	10_{216} (22)	8_{168} (23)	
L: $+8 \rightarrow 248$	L: $+8 \rightarrow 272$	P: $+4 \rightarrow 244$	P: $+4 \rightarrow 220$	P: $+4 \rightarrow 172$	
D: $+16 \rightarrow 256$	D: $+16 \rightarrow 280$				
R: $+12 \rightarrow 252$	R: $+12 \rightarrow 276$				
	P: $+4 \rightarrow 268$				

Therefore, during the traversal, the application accesses the following memory positions: 176, 80, 32, 8, 16, 12, 4, 40, 36, 56, 64, 60, 52, 28, 88, 84, 128, 104, 112, 108, 100, 136, 132, 152, 160, 156, 148, 124, 76, 184, 180, 224, 200, 208, 204, 196, 232, 228, 248, 256, 252, 272, 280, 276, 268, 244, 220 and 172. This access pattern does not exhibit an easily recognizable form of spatial locality.

As the second tree operation, consider the retrieval of the data element corresponding to the key "5." The application visits the nodes "8," "4," and "6," accessing the memory addresses 168, 176, 72, 84, 120, 128, 96 and 112:

8_{168} (1)	4_{72} (2)	6_{120} (3)	5_{96} (4)
K: +0 → 168	K: +0 → 72	K: +0 → 120	K: +0 → 96
L: +8 → 176	R: +12 → 84	L: +8 → 128	D: +16 → 112

Finally, the insertion of the last node, "12," requires the following accesses, not counting the balance calculation in the branch up to "4" and the accesses required to perform the rotation at the root level:

4_{72} (1)	8_{168} (2)	10_{216} (3)	11_{240} (4)
K: +0 → 72	K: +0 → 168	K: +0 → 216	K: +0 → 240
R: +12 → 84	R: +12 → 180	R: +12 → 228	R: +12 → 252
12_{264} (5)			
K: +0 → 264			
P: +4 → 268			
L: +8 → 272			
R: +12 → 276			
D: +16 → 280			
Balance: +20 → 284			

For the accesses included in the table, the visited memory addresses are: 72, 84, 168, 180, 216, 228, 240, 252, 264, 268, 272, 276, 280 and 284.

Up to now, we have considered the memory accesses executed by the application only to get a sense of the effect of the intricacies of DDTs in their order. However, we can also do a few quick estimations of the cost associated with them using different memory elements. First, let us consider a simple system with a cache memory, its degree of associativity being irrelevant as long as it has a capacity of at least 512 B. For the cache line size, we can explore the cases of 4 and 16 words per cache line. We can calculate the number of accesses to the main DRAM executed during the insertion of element "12," which involves 14 words of memory, starting from an empty cache condition:

(a) Lines of 16 words. Cache lines accessed: 1, 1, 2, 2, 3, 3, 3, 3, 4, 4, 4, 4, 4 and 4.
   4 different lines accessed. 4 lines × 16 words = 64 accesses to the DRAM.
   Overhead: (64 − 14) / 14 = 3.57.
(b) Lines of 4 words. Cache lines accessed: 4, 5, 10, 11, 13, 14, 15, 15, 16, 16, 17, 17, 17 and 17.

9 different lines accessed. 9 lines × 4 words = 36 accesses to the DRAM.
Overhead: $(36 - 14) / 14 = 1.57$.

Second, consider a system with a scratchpad (SRAM) memory. As every word in the memory is independent, the processor accesses only those positions referenced by the application. Only 14 memory accesses are performed, in 14 memory cycles, with no energy or latency overheads (i.e., overhead is 1.0). Finally, a system with only SDRAM would also execute only 14 accesses, leaving aside the requirements to open the appropriate DRAM row, and assuming no row conflicts.[2]

Although these calculations are oversimplified estimations, the relevant factor is that the systems without cache memory do not waste energy in unneeded operations in cases such as this one. In this simple example, cache performance would improve drastically with further high temporal locality traversals as many accesses would be served without accessing the external DRAM. However, the effect is not negligible when the number of nodes in the tree increases and the logical connections become more scattered over the memory space, with every operation accessing different subsets of nodes. Additionally, the mixed pattern of allocations and deallocations will cause over time a dispersion of the addresses assigned to related nodes. Even worse, the application will probably interleave accesses to the tree with accesses to other data structures, or the cache may be shared with other threads. The additional pressure over the cache will force more evictions and more transfers between levels, exacerbating the ill effects of accessing, transferring, and storing data words that are not going to be used. Finally, cache efficiency suffers also from the fact that cache line size and object size do not necessarily match. Common options are padding (area and energy waste in storing and transferring filler words) or sharing cache lines with other objects (energy waste transferring words not used, possible false sharing phenomena in multiprocessor environments [5, p. 476]).

3.2.3   Another AVL Example, with Random Data

The elements that we inserted in the AVL tree during the previous example were already ordered according to their keys, which forced many rotations in the tree. However, adding elements in random order produces a similar scattering of logical nodes in memory addresses because although less rotations are performed, randomly inserted keys go randomly towards left or right children. In the end, nodes that were created in contiguous memory addresses (assuming again a simplistic DMM model) become logically linked to nodes in very different addresses.

---

[2]An on-chip scratchpad (SRAM) or cache memory usually works at a higher frequency than an external DRAM chip. Therefore, in this discussion we refer to memory cycles in contraposition to processor cycles or real time.

Let us repeat the AVL experiment with twelve random numbers: 491, 73, 497, 564, 509, 680, 136, 963, 273, 12, 514 and 825. The final configuration of the tree becomes:

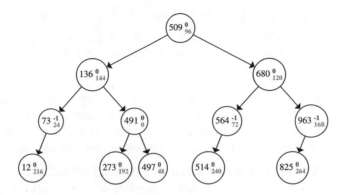

And the layout of the tree's nodes in memory is:

491 ₀	73 ₂₄	497 ₄₈	564 ₇₂	509 ₉₆	680 ₁₂₀
136 ₁₄₄	963 ₁₆₈	273 ₁₉₂	12 ₂₁₆	514 ₂₄₀	825 ₂₆₄

In order to get a better picture of the spreading of logical nodes over physical cache lines, we can color the previous graph assigning a different color to each cache line. The nodes arrive consecutively and are assigned successive memory addresses, thus using consecutive cache lines. However, the position that they occupy in the logical structure is very different and can potentially change along the DDT lifetime.

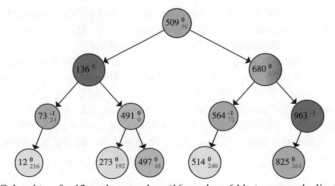

Colored tree for 12 random numbers (16 words or 64 bytes per cache line).

Given the memory layout that we have calculated for our AVL tree, a fetch of the data associated to the key value "273" requires visiting the nodes "509," "136," and "491" with the following access pattern:

509_{96} [1]	136_{144} [2]	491_0 [3]	273_{192} [4]
K: $+0 \to 96$	K: $+0 \to 144$	K: $+0 \to 0$	K: $+0 \to 192$
L: $+8 \to 104$	R: $+12 \to 156$	L: $+8 \to 8$	D: $+16 \to 208$

According to this pattern, the application reads memory addresses in the following sequence: 96, 104, 144, 156, 0, 8, 192 and 208. Making similar assumptions to that in the previous example, we can calculate the number of accesses to the main DRAM for several cache configurations and reach similar conclusions:

(a) Lines of 16 words. Cache lines accessed: 1, 1, 2, 2, 0, 0, 3 and 3.
   4 different lines accessed. 4 lines × 16 words = 64 accesses to the DRAM. Overhead: $(64 - 8) / 8 = 7$.
(b) Lines of 4 words. Cache lines accessed: 6, 6, 9, 9, 0, 0, 12 and 13. 5 different lines accessed. 5 lines × 4 words = 20 accesses to the DRAM. Overhead: $(20 - 8) / 8 = 1.5$.

3.3　Summary: DDTs Can Reduce Access Locality

Cache memories rely on the exploitation of the locality properties of data accesses by means of prefetching and storing recently used data. However, the use of DDTs creates important issues because *logically adjacent linked nodes are not necessarily stored in consecutive memory addresses* (the DMM may serve successive requests with unrelated memory blocks and nodes may be added and deleted at any position in a dynamically linked structure), breaking the spatial locality assumption, and, in some structures such as trees, *the path taken can be very different from one traversal to the next one*, thus hindering also the temporal locality.

These considerations support the thesis defended in this text: Embedded systems with energy or timing constraints whose software applications have a low data access locality due to the use of DDTs, which is especially common in object-oriented languages, and in general with applications applying virtualization, should be designed considering the utilization of explicitly addressable (i.e., non-transparent) memories rather than caches. In that scenario the placement problem becomes relevant.

References

1. Adel'son-Vel'skii, G.M., Landis, E.M.: An algorithm for the organization of information. Sov. Math. Dokl. **3**, 1259–1263 (1962). http://monet.skku.ac.kr/course_materials/undergraduate/al/lecture/2006/avl.pdf
2. Brassard, G., Bratley, T.: Fundamentals of Algorithmics, 1st (Spanish) edn, pp. 227–230. Prentice Hall, Englewood Cliffs (1996)
3. Chilimbi, T.M., Davidson, B., Larus, J.R.: Cache-conscious structure definition. In: Proceedings of the ACM SIGPLAN Conference on Programming Language Design and Implementation (PLDI), pp. 13–24. ACM Press, Atlanta (1999). https://doi.org/10.1145/301618.301635
4. Daylight, E., Atienza, D., Vandecappelle, A., Catthoor, F., Mendías, J.M.: Memory-access-aware data structure transformations for embedded software with dynamic data accesses. IEEE Trans. Very Large Scale Integr. Syst. **12**(3), 269–280 (2004). https://doi.org/10.1109/TVLSI.2004.824303
5. Herlihy, M., Shavit, N.: The Art of Multiprocessor Programming, 1st (revised) edn. Morgan Kaufmann, Waltham (2012)
6. Marchal, P., Catthoor, F., Bruni, D., Benini, L., Gómez, J.I., Piñuel, L.: Integrated task scheduling and data assignment for SDRAMs in dynamic applications. IEEE Design Test Comput. **21**(5), 378–387 (2004). https://doi.org/10.1109/MDT.2004.66
7. Weiss, M.A.: Estructuras de Datos y Algoritmos, 1st (Spanish) edn., p. 489. Addison-Wesley, Reading (1995)

Chapter 4
Methodology for the Placement of Dynamic Data Objects

To palliate the consequences of the speed disparity between memories and processors, computer architects introduced the idea of combining small and fast memories with bigger—albeit slower—ones, effectively creating a memory hierarchy. The subsequent problem of data placement—which data objects should reside in each memory—was solved in an almost transparent way with the introduction of the cache memory. However, the widespread use of dynamic memory can hinder the main property underlaying the good performance of caches and similar techniques: Data access locality. Even though a good use of prefetching can reduce the impact on performance, the increase on energy consumption due to futile data movements is more difficult to conceal. Figure 4.1 summarizes this situation and the solutions proposed in this work.

During the early days of our work, someone suggested that the placement of dynamic objects could be trivially reduced to the placement of static objects—on which lots of work had already been invested: "Simply give us the pool that holds all your dynamic objects as if it were an array with given access characteristics and our tools for the placement of static objects will place it as an additional static array." That proposal encloses a significant simplification: All the dynamic objects are considered as a whole, without distinguishing those heavily accessed from those seldom used. The result is likely a poor placement of dynamic data objects and improvable performance, justifying the development of specific techniques.

The key to the problem is discriminating between dynamic objects with different characteristics instead of mixing them in pools and treating them all as a single big array. In other words, the problem lies in differentiating between dynamic data types (DDTs) before putting them into pools, rather than in the placement of the resulting pools themselves, which is a more studied problem.

Our proposal is based on avoiding data movements between elements in the memory subsystem using the dynamic memory manager (DMM) to produce a careful placement of dynamic data objects on memory resources. But, how is it possible to achieve such a good placement? What are the new responsibilities

© Springer Nature Switzerland AG 2020

M. Peón Quirós et al., *Heterogeneous Memory Organizations in Embedded Systems*, https://doi.org/10.1007/978-3-030-37432-7_4

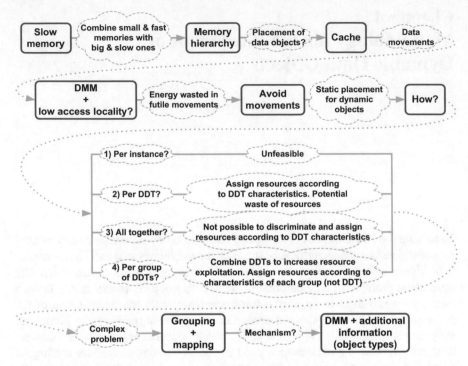

Fig. 4.1 A careful placement of dynamic data objects can improve the performance of the memory subsystem, avoiding the impact of low access locality on cache memories. To cope with the problem's complexity and improve resource exploitation, we propose a two-step process that groups dynamic data types and uses the dynamic memory manager to implement a static placement of dynamic data objects

of the dynamic memory manager? In this chapter we will explore the available options to build a dynamic memory manager with support for data placement and propose simple solutions to the main problems. The outcome will be a methodology encompassing the whole design process—from application characterization to deployment.

Definition Through the rest of this chapter, we shall use the term "heap" to denote the address range, "dynamic memory manager" (DMM) for the algorithms, and "pool" for the algorithms, their internal control data structures and the address range as a whole.

4.1   Choices in Data Placement

This section discusses briefly the different types of data placement and the granularity levels at which it can be implemented. It presents also the main requisites to achieve an efficient data placement.

4.1.1 The Musts of Data Placement

Four are the fundamental requisites for a data placement technique to achieve an efficient use of resources:

1. The most accessed data must be in the most efficient resources—or most of the accesses must be to the most efficient memory resources. Caching techniques rely on the property of access locality. Instead, our proposal uses information extracted through profiling to identify the most accessed data objects.
2. Recycling of resources. Caching techniques achieve it through data movements. Our methodology uses the grouping step to identify and place together objects that alternate in the use of resources.
3. Avoiding unnecessary overheads. For example, moving data that will not be used (such as prefetching of long or additional cache lines) or with low reutilization. Caching techniques rely on access locality; software-based ones can use knowledge on the application to avoid some useless movements. Our methodology produces a static placement of pools with no overheads in terms of data movements.
4. Avoiding displacement of very accessed data so that they have to be brought back immediately. For example, the victim cache was introduced to palliate this problem with direct-mapped caches [16]. Our methodology places data objects according to their characteristics, ensuring that seldom accessed objects are not migrated to closer resources; thus, they cannot displace other objects.

4.1.2 Types of Data Placement

Data placement in general can be approached in several ways depending on the nature of the data objects and the properties of the desired solutions. Table 4.1 shows the main options.

Table 4.1 Data placement according to implementation time and object nature

	Static data	Dynamic data
Static	Direct placement per object	Direct placement at allocation time
Placement	Problem: low exploitation	Problem: resource underuse
Dynamic	Cache. SW-based movements	Cache. SW-based movements
Placement		Problem: low access locality

Static placement options choose a fixed position in the memory subsystem during object creation. Dynamic placement options may move data at run-time to bring closer those that are likely to be more accessed over the next period. Data movements may happen at the object or byte levels

4.1.2.1   Placement Policy

A static placement policy decides explicitly the memory position that every data object will have during the whole execution. The most used data objects can be placed in the most efficient resources, but accesses to the remaining ones bear higher costs. This happens even when the objects placed in the efficient memories are not being used—even if they hold no valid data at that time—because they have been granted exclusive ownership of their resources.

On the contrary, a dynamic placement policy can move data objects (or portions thereof) to keep track of changes in the application access patterns and obtain a better exploitation of resources. This is especially effective when the moved data are reused, so that the cost of the movement is shared among several subsequent accesses. Techniques such as cache memories or software prefetching belong to that category. Whereas static placement usually requires less chip area and energy [5], dynamic placement offers better adaptability for applications that alternate between phases with different behavior or changing working conditions.

4.1.2.2   Data Nature

Static data objects cannot be broken over non-contiguous memory resources because compilers commonly assume that all the fields in a structure are consecutive and generate code to access them as an offset from a base address. Therefore, static placement techniques [17, 18, 25–27] cannot use a greedy algorithm to calculate (optimally) the placement; hence, they resort to dynamic programming or integer linear programming (ILP) to produce optimal results. Although those static placement techniques can still be employed to guarantee that critical data objects are always accessed with a minimum delay, they easily become impractical for large numbers of objects, which is precisely the situation with dynamic data. Anyways, the placement of dynamic data objects cannot be fully computed at design time because their numbers and sizes are unknown until run-time. Consider, for example, that every location at the source code where a dynamic object is created can be executed multiple times and that the life span of the objects can extend further than the scope at which they were created.[1]

Dynamic placement techniques based on data movements (for caching and prefetching) may also be inadequate because the very nature of dynamic data objects reduces access locality and thus their efficiency. They would still cope efficiently, for instance, with algorithms that operate on dynamically allocated vectors (e.g., a dynamically allocated buffer to hold a video frame of variable size), but not so much with traversals of linked data structures.

---

[1] A technique to verify whether this is the case is escape analysis. However, a simple example is when dynamic objects are added to a container created out of the current scope.

4.1.3 Granularity of Data Placement

Whereas the previous paragraphs looked into data placement in general, here we analyze specifically the dynamic data placement problem (placing all the instances of all the DDTs of the application into the memory elements) and the different granularity levels at which it can be tackled. For instance, it can be solved at the level of each individual instance, albeit at a high expense because their numbers are unknown and they are created and destroyed along time, which may render some early decisions suboptimal. The opposite direction would be to group all the DDTs into a single pool, but that would be quite close to the standard behavior of general-purpose DMMs without placement constraints.

The next option that appears intuitively is to characterize the mean behavior of the instances of each DDT and assign memory resources to the DDT itself.[2] In that way placement decisions can be taken for each DDT independently, while keeping the complexity bounded: Space is assigned to the pool corresponding to each DDT once at the beginning of the execution, and classic dynamic memory management techniques are used to administer the space inside the pool. However, avoiding data movements across elements in the memory hierarchy presents the risk of high resource underexploitation when there are few alive instances of the DDT assigned to a resource. Here, we will explore a mechanism that tries to compensate for it by grouping together DDTs that have complementary footprint demands along time, so that "valleys" from ones compensate for "peaks" of the others and the overall exploitation ratio of the memory resources is kept as high as possible during the whole execution time. The difficulty of grouping lies then on identifying DDTs whose instances have similar access frequencies and patterns, so that any of these instances equally benefit from the assigned resources.

The following sections describe each of the options mentioned before in more depth:

4.1.3.1   All DDTs Together

The DDTs are put in a single pool, which uses all the available memory resources. Heap space can be reused for instances of any DDT indistinctly through splitting and coalescing. This is the placement option that has the smallest footprint, assuming that the selected DMM is able to manage all the different sizes and allocation patterns without incurring a high fragmentation. However, as the DMM does not differentiate between the instances of each DDT, new instances of very accessed DDTs would have to be placed in less efficient resources when the better ones are already occupied—perhaps by much less accessed instances. The DMM can try to assign resources according to the size of the objects, but unfortunately different

---

[2]It may happen that the instances of a DDT have different access characteristics. A mechanism to differentiate between them would be an interesting basis for future work.

DDTs with instances of the same size will often have a quite different number of accesses, limiting the effectiveness of such placement: Frequently and seldom accessed objects of the same size would all use the same memory resources.

This placement is what a classic all-purpose DMM would implement, which is precisely the situation that we want to avoid: The DMM is concerned only with reducing overhead (internal data structures and fragmentation) and finding a suitable block efficiently. An additional drawback of this option for the placement of linked data structures is that the DMM can choose any block from any position in the heap without constraints regarding the memory resource or position where it comes from. Thus, depending on the DMM's policy for reusing blocks (e.g., LIFO), nodes in a linked structure can end up in very different places of the memory map, reducing access locality.

4.1.3.2 One Pool Per DDT

This option is the opposite of the previous one: Each DDT gets its own pool that can be tailored to its needs. Placement can select the most appropriate memory resource for the instances of that DDT, without interferences from other ones. On the downside, memory wastage can increase significantly: Even when there are few instances—or none at all—of the DDT that is assigned to an efficient memory resource that space cannot be used for instances of other DDTs. This is an important issue because movement-based caching techniques can exploit those resources, especially with higher degrees of associativity, by temporarily moving the data that are being used at each moment into the available resources—their problem in this respect is not wasting space, but seldom accessed objects evicting frequently accessed ones.

A possibility to palliate this problem could be temporarily overriding placement decisions to create some instances in the unused space. However, that would simply delay the problem until the moment when the space is needed again for instances of the assigned DDT. What should be done then? It would be easy to end up designing solutions equivalent to the very same caching schemes that we were trying to avoid, but without their transparency.

Just for the sake of this discussion, let us examine briefly some of the options that could be considered to face this issue:

- Data migration at page granularity. It can be implemented in systems that support virtual memory for pools bigger than the size of a hardware page (e.g., 4 KB). The pools are divided into page-sized blocks. As the DMM starts assigning space in the pool, it tries to use space from the minimum number of pages, in a similar way to the old "wilderness preservation heuristic" [28, pp. 33–34]. If other pool has (whole) unused pages in a more efficient resource, the system maps the pool's new pages in that resource. When the pool that owns the efficient memory resource needs to use that page for its own objects, the whole page is migrated to the correct memory resource.

In principle, this idea seems to allow some pools to exploit better available resources while they are underused by the DDTs that own them. However, as objects become scattered and free space fragmented (assuming that it cannot always be coalesced) the possibility of exploiting whole pages becomes more limited, perhaps making this mechanism suitable only for the first stages of the application execution. Future research can delve deeper into this issue.

- Object migration. Techniques to patch references (pointers) such as the ones used by garbage collectors might be used to migrate individual objects once the borrowed space is claimed by the owner DDT. Future research may evaluate the trade-off between the advantages obtained by exploiting the unused space for short-lived objects that are never actually migrated, and the cost of migrating and fixing references to long-lived objects.[3]
- Double indirection. With support from the compiler, this mechanism could help to migrate some objects without more complex techniques. Considering that it would affect only objects with lower amounts of accesses, the impact on performance might be limited under favorable access patterns. The trade-off between reloading the reference before every access or locking the objects in multithreaded environments would also need to be explored.

All of these options involve some form of data migration. A final consideration is that if objects are placed in a different pool (except for the case of migration at the page level), that pool needs to be general enough as to accommodate objects of other sizes, which partially defeats the whole purpose of DMM specialization. As none of the previous options is entirely satisfying, it seems clear that a completely different approach is needed to make static placement a viable option for dynamic data objects.

4.1.3.3 Per Individual Instance

This granularity level represents the ultimate placement option: Choosing the resource where to place each individual data object as it is created. However, contrary to the situation with static data, the number and characteristics of individual dynamic data instances cannot usually be foreseen: The number of instances actually created at a given code location may be unknown (for static data, that number is one). In order to correctly place each object, the DMM would need to know the access characteristics of that specific instance, but it seems currently difficult to devise any mechanism to obtain that information, except perhaps for very domain-specific cases. Therefore, we will deem this option as unfeasible for practical purposes in the scope of the techniques presented here.

---

[3] We believe that the grouping mechanism proposed in this work identifies such short-lived DDTs and automatically exploits any available space from other pools; hence, it voids the need for migrating this type of objects.

An interesting consideration is that, as the exact number of instances cannot be calculated until each one is effectively created, the system has to take each placement decision without knowledge about future events. Thus, when a new instance is created, the situation may not be optimal. Had the system known that the new object was indeed going to be created, it could have decided to save a resource instead of assigning it to a previously created one. In the absence of a mechanism to "undo" previous decisions, placement can be suboptimal. Maybe future systems will allow identifying the number of accesses to the objects and migrate them accordingly to adjust their placement a posteriori.

4.1.3.4 By Groups of DDTs

Putting together "compatible" DDTs in the same pool. This last option is intriguing: How are the DDTs in each group selected? And, why group them at all? Although the number of dynamic data instances created by an application depends on conditions determined at run-time, oftentimes the designer can use profiling or other techniques to identify patterns of behavior. Particularly, it may be possible to obtain typical footprint and access profiles for the DDTs and group those ones that present similar characteristics or whose instances are mostly alive at complementary times—so that either they can share the same space at different times or it does not really matter whose instances use the assigned resources. The advantage of this option is that dissimilar DDTs can be treated separately while the total memory requirements are kept under control. In other words, grouping of DDTs for placement helps in provisioning dedicated space for the instances of the most accessed ones while improving resource utilization.

We propose this option as a mechanism to assign dedicated space to some DDTs (static placement) while aiming at a high resource exploitation.

4.2 Data Placement from a Theoretical Perspective

Computational complexity is an evolving theoretical field with important consequences on the practice of computing engineering. We invite interested readers to delve deeper into it with the classic textbook from Cormen et al. [13, Chap. 34–35]. In this section we will take just a glimpse of the world of computational complexity, introducing the concepts that can help to understand the complexity of the data placement problem.

4.2.1   Notes on Computational Complexity

In the context of computational complexity, problems that can be solved exactly with an algorithm that executes in polynomial time in the worst case are said to be in the P complexity class. Those problems are generally regarded as "solvable," although a problem with a complexity in the order of $O(n^{100})$ is hardly easy to solve—anyways, typical examples are in the order of $O(n^2)$ or $O(n^3)$ at most, where n is the size of the problem. A very important property of P is that its algorithms can be composed and the result is still in P.

Problems for which no exact algorithm working in polynomial time has ever been devised, but whose solutions—if given by, say, an oracle—can be verified in polynomial time by a deterministic machine are said to be in the NP class. An interesting remark is that not knowing any algorithm to solve a problem in polynomial time is not the same than being sure that such an algorithm does not exist—and can thus never be found. In fact, we know that $P \subseteq NP$, but the question of whether $P = NP$ is the holy grail of computational complexity: On May 24, 2000, the Clay Mathematics Institute of Cambridge announced a one million dollar prize for the person who can solve that riddle [10] as formalized by Cook [12].

NP-complete is a special class of problems in NP. A problem p is said to be in NP-complete if every other problem in NP can be polynomially reduced to it, that is, if there is a transformation working in polynomial time that adapts the inputs to the other problem into inputs to p and another transformation that converts the solution of p into a solution for the original problem. Therefore, we can informally say that NP-complete contains the hardest problems in NP.

> [...] This class [NP-complete] has the surprising property that if any NP-complete problem can be solved in polynomial time, then every problem in NP has a polynomial-time solution, that is, $P = NP$. [13, Chap. 34.3]

Strictly speaking, NP-complete contains decision problems, that is, problems that admit "yes" or "no" as an answer. However, we are commonly concerned with problems that require more complex solutions. A common technique is to define the decision problem related to a more general one, so that if the decision problem is shown to be in NP-complete, the general problem is then said to be in the class of NP-hard problems. This transformation can be done, for instance, changing a knapsack problem into a question such as "Can a subset of objects be selected to fill no more than v volume units and with a minimum of b benefit units?"

The difficulty in solving some problems has motivated the development of a complete theory of approximation algorithms. Although many problems are intractable in the worst case, many can be approximated within a determined bound with efficient algorithms. The approximation factor is usually stated as $\epsilon > 0$, so that a typical approximation algorithm will find a solution within a $(1 \pm \epsilon)$ factor of the optimal—the sign depending on whether the problem is a maximization or minimization one. A useful class of problems is FPTAS (fully polynomial-time approximation scheme), which is defined as the set of problems that can be approximated with an algorithm bounded in time both by $1/\epsilon$ and the problem size.

However, an FPTAS cannot be found for the most complex instances of the knapsack family of problems unless $\mathcal{P} = \mathcal{NP}$ [21, pp. 22–23, for all this paragraph]. The most complex ones are said to be hard even to approximate. For example:

> The Multiple Knapsack Problem is $\mathcal{NP}$-hard in the strong sense, and thus any dynamic programming approach would result in strictly exponential time bounds. Most of the literature has thus been focused on branch-and-bound techniques [...] [21, p. 172]

More recent works, such as the one by Chekuri and Khanna [9], propose that the multiple knapsack problem is indeed the most complex special case of GAP that is not APX-hard—i.e., that is not "hard" even to approximate.

4.2.2   Data Placement

The problem of placement for dynamic data objects on heterogeneous memory subsystems is complex to solve. Although a formal proof would require more technical knowledge than we currently possess, we will assume that it is a generalization of the (minimization) general assignment problem (GAP). In this section we will compare briefly both problems.

The efforts of theorists and practitioners in computational complexity have been long elicited by a family of problems of which perhaps the simplest is the 0/1 knapsack, in which a set of indivisible objects, each with its own intrinsic value, needs to be fit in a knapsack of limited capacity. The goal is to maximize the value of the chosen objects.[4] The problem can be complicated in several ways. For example, the multiple knapsack problem has several containers to fill, maximizing the aggregate value. Bin packing has the goal of packing all the objects in a set in the minimum possible amount of containers. The multiple choice knapsack problem divides the objects in different classes from which a maximum number of items can be taken.

Similar problems include also instances of scheduling, where some tasks need to be scheduled on a number of processors (or orders assigned to production lines) to minimize the total execution time. The complexity of the problem increases if the processors have distinct characteristics so that the cost of each task depends on the processor that executes it. An interesting variation is the virtual machine (VM) colocation problem because the size of each VM can vary depending on the rest of machines that are assigned to the same physical server: Quite frequently pages from several VMs will have identical contents and the hypervisor will be able to serve them all with a single physical page [23]. The particularity of this problem is

---

[4]Fractional or continuous knapsack, where the objects can be split at any point (e.g., liquids), is sometimes regarded as a different type of problem. However, that problem is also interesting because it can be exactly solved by a greedy algorithm in logarithmic time $O(n \log n)$ using sorting or even in linear time $O(n)$ using weighted medians. More importantly, it can be used to quickly obtain upper bounds in branch-and-bound schemes to solve the harder versions [21, p. 18].

that the set of previously selected objects affects the size or cost of the remaining ones. Although all the knapsack problems (except the continuous one) belong to the category of $\mathcal{NP}$-hard problems—they are "hard" to solve in the worst case—researchers have been able to devise techniques to solve or approximate many cases of practical interest in polynomial time, often in less than 1 s [21, pp. 9–10].

The general assignment problem (GAP) raises the complexity level even more, because it has multiple containers and the cost and benefit of each object depends on the container into which it is assigned:

Instance: A pair (B, S) where B is a set of M bins and S is a set of N items. Each bin $c_j \in B$ has capacity $c(j)$, and for each item i and bin c_j we are given a size $s(i, j)$ and a profit $p(i, j)$. [11]

In this work, we assume that tackling the problem of placement at the level of individual dynamic data objects is unfeasible and therefore we propose to approach it as the placement of DDTs. However, in this form the problem has still more degrees of freedom than other problems from the same family because the number and size of the containers is not fixed. It consists on assigning a set of DDTs to a set of memory resources, without exceeding the capacity of each resource and minimizing the total cost of the application accesses to the data objects. As in the GAP problem, multiple memory resources (containers) exist, each with a different capacity, and the cost of accessing each DDT is different according to the characteristics of each module. However, the size of the containers themselves can vary as well, adjusting to the combined, not added, size of the DDTs that they contain. As in the VM-colocation problem, the size of each DDT depends on the other DDTs (objects) that have been already selected in that resource.[5] Furthermore, for some memory resources, the very cost of accessing a DDT may depend on the other DDTs placed there. That is, for instance, the case of objects assigned to the same bank of a DRAM where accesses to each one can create row misses to access the others.

Data placement presents an additional difficulty. Hard instances of the previous problems are usually solved with branch-and-bound techniques. To prune the search space and avoid a full exploration, they require a mechanism to calculate an upper bound (for minimization) on the cost of the current partial solution. However, in the data placement problem, assessing the cost of a partial solution can be quite difficult because it depends on the way that the application uses the placed data objects. Ideally, and assuming that the memory traces obtained during profiling are sufficiently representative, a simulator like the one explained with our methodology could be used to calculate the exact cost of a complete placement solution. That approach is more difficult for partial solutions, though, as the cost of accesses to data objects belonging to DDTs not yet placed cannot be easily calculated. An option

---

[5]In this text we propose to group DDTs with similar characteristics to overcome the inefficiencies in resource exploitation of a static data placement. The size of a group depends on the specific objects in it, hence the similarity with the VM-colocation problem. Memory fragmentation inside the pools also contributes to this effect.

that would give a very coarse upper bound is to assume that all non-placed DDTs are placed in main memory.

Another issue is that the simulation of whole memory traces, although a fast process, may require a significant amount of time, especially if the estimation has to be calculated for many different nodes during the search process. One alternative could be using high-level estimators to produce a rough approximation of the cost of a solution. Those estimators would simply multiply the total number of accesses to the instances of each DDT by the cost of each access to the memory module where they are (tentatively) placed. For DDTs placed in SRAMs, the estimation should be pretty close to the real cost. For those placed on DRAMs, on the contrary, the estimation can deviate significantly from the real value as interactions between accesses to different DDTs in the banks of a DRAM can force an indeterminate number of row misses, with the corresponding increase in energy consumption and access time. In any case, we leave the exploration of such possibilities to further work.

4.3   A Practical Proposal: Grouping and Mapping

The placement of dynamic data objects is a complex problem that we believe to belong to the family of the (minimization) general assignment problem (GAP). Therefore, we assume that tackling the problem at the level of individual dynamic data objects is unfeasible and propose to approach it as the placement of DDTs. However, this variant of the problem is still complex with respect to other problems from the same family because the number and size of the containers are not fixed. Thus, it is possible to put one object (DDT) in each container, but also to combine several DDTs in the same one and then both the size (memory footprint) of the DDTs (because of space reutilization similar to the case of VM-colocation [23]) and the size of the container itself (which adjusts to the combined, not added, size of the DDTs) vary. Moreover, the cost of the accesses to a DDT (i.e., the cost of the object) depends not only on the container (memory resource) where it is placed, but the presence of other DDTs in the same container can modify it, as is the case of DDTs placed in the same bank of a DRAM.

The proposal that we are going to study here addresses the complexity of the problem in two ways. First, it raises the level of abstraction: Instead of working with individual data objects or individual DDTs, it identifies groups of DDTs with similar characteristics and includes them in the same pool of dynamic memory. This step is the key to improve resource exploitation. Whereas pools are the entities considered during placement, individual objects are allocated in the pools at run-time by a dynamic memory manager. In this way, this proposal implements a static placement of pools that reduces the risk of fruitless data movements while observing the dynamics of object creation and destruction.

Second, we propose to break the original problem into two steps that are solved independently: The aforementioned classification of DDTs that can share a resource into groups, and the placement of those groups over the memory resources (Fig. 4.2). This constitutes in proper terms a heuristic akin to the ones used to solve other complex forms of the assignment family of problems. Although not guaranteeing optimal solutions, it achieves very good results when compared with traditional options, as shown in the experiments of Chap. 6. Additionally, most of the hard work can be done at design time, limiting hardware support or computations during run-time. The outcome is a process that provisions dedicated space for the instances of the most accessed DDTs, while keeping resource utilization high.

The grouping step analyzes the DDTs of the application and creates groups according to their access characteristics and the evolution of their footprint along execution time. It is a platform-independent step that builds groups according only to the characteristics of the DDTs and the values assigned by the designer to a set of configurable parameters, enabling a generic grouping of DDTs at design time, while exact resources are assigned later. Contrary to problems such as bin packing, the grouping algorithm does not impose a limit on the number or size of the containers (groups): If the designer forces a limit on their number, the DDTs that are not accepted into any group are simply pushed into the last one.

The second step, mapping, is concerned with the correspondence of divisible objects (pools) over a (reasonable) number of containers, without caring for the interactions among the entities that form each object. Thus, the mapping step assigns memory resources to each group using a fractional knapsack formulation.[6] This opens the door for a run-time implementation that places the groups according to the

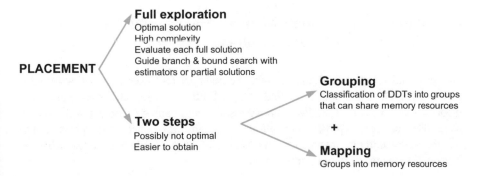

Fig. 4.2 Our methodology proposes as a heuristic splitting the original placement problem into two parts that are solved independently. The first one consists on identifying DDTs that can share the same resources; the second assigns the available memory resources to those groups

---

[6]Increasing the abstraction level at which a problem is solved may introduce some inefficiencies, but it offers other advantages. For example, whereas individual variables cannot be broken over non-contiguous memory resources, dynamic memory pools can—an object allocated in a pool cannot use a memory block that spawns over two disjoint address ranges, though, but this can be solved by the DMM at the cost of possibly slightly higher fragmentation. In this way, the placement of divisible pools can be solved optimally with a greedy algorithm.

resources available in the exact moment of execution and for that concrete system considering, for instance, the possibility of different instantiations of a general platform or a graceful reduction of performance due to system degradation along its lifetime.

This scheme works well when the application has DDTs with complementary footprints or if the number of instances extant at a given time of each DDT is close to the maximum footprint of the DDT. In other cases, compatible DDTs may not be found; hence, the exploitation ratio of the memory resources will decrease and performance will suffer. In comparison with other techniques such as hardware caches, a part of the resources might then be underused. Whether the explicit placement implemented through grouping and mapping is more energy or performance efficient than a cache-based memory hierarchy in those situations or not—avoiding data movements under low-locality conditions can still be more efficient even if the resources are not fully exploited all the time—is something that probably has to be evaluated on a case-by-case basis.

It is also interesting to remark that, as the break up into the steps of grouping and mapping is a form of heuristic to cope with the complexity of placement, so are the specific algorithms that we propose to implement each of these steps. For example, to limit the cost of finding solutions, the algorithm used here for grouping uses a straightforward method with some parameters that the designer can use to refine the solutions found. Future research may identify better options for each of the algorithms that we propose in this chapter.

4.4   Designing a Dynamic Memory Manager for Data Placement with DDT Grouping

Our proposal relies on a dynamic memory manager that uses knowledge on allocation sizes, DDT characteristics and compatibilities between DDTs (i.e., grouping information) to assign specific memory resources to each data object at allocation time. This section explores how all these capabilities can be incorporated into the dynamic memory manager.[7] The final goal is to supply the DMM with the information required for data placement:

- DDTs that can be grouped.
- Correspondence between DDTs/pools and memory resources.
- The data type of each object that is allocated.

Figure 4.3 presents the choices during the construction of a DMM that concern specifically data placement. The decision trees shown in the figure are independent from each other: It is possible to pick an option at one of the trees and combine

---

[7]A comprehensive analysis of the full design space for dynamic memory managers, without data placement, was presented by Atienza et al. [1, 3].

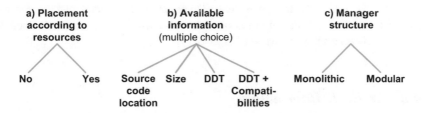

Fig. 4.3 Decisions relevant for data placement during the design of a dynamic memory manager: (**a**) Has the DMM to include data placement capabilities? (**b**) What kind of information is available for the DMM at allocation time? (**c**) Does the DMM take global decisions or is it a modular collection of particularly optimized DMMs?

it with any other decisions from the rest of them. However, the decision taken at one tree can influence the implementation of the decisions taken in the next ones. We elaborate more on the ordering of choices in Sect. 4.4.4 after all of them are presented.

4.4.1   Placement According to Memory Resources

The first decision, it may seem obvious in this context, is whether the dynamic memory manager should implement data placement or not (Fig. 4.3a). That is, whether the DMM has to take into account the characteristics of the memory resources underlaying the available memory blocks and the object that will be created, or ignore them and treat all the blocks as equal in this regard as is the most frequent case with general-purpose managers for desktop systems.

If the DMM does not have to care about data placement, its focus will be mainly to reduce wasted space due to fragmentation (similar to the scraps produced while covering a surface with pieces of restricted sizes), time required to serve each request (usually linked to the number of memory accesses performed by the manager to find the right block) and the overhead introduced by the manager itself with its internal data structures. Those structures are usually built inside the heap taking advantage of the space in free blocks themselves, but frequently require also some space for headers or footers in used blocks (e.g., the size of a used block may be stored in front of the allocated space) and thus impose a lower limit on the size of allocated blocks. Depending on the available information, the memory manager may rely on independent lists for blocks of different sizes or even different heaps for each DDT. Even if the DMM does not produce the data placement, other software techniques such as array blocking or prefetching can still be used for dynamic objects with predictable access patterns such as dynamically allocated vectors.

If the duties of the DMM include data placement, it has to place each object according to its size and number and pattern of accesses. The different granularity levels at which placement can be implemented have already been discussed in Sect. 4.1.3; the conclusion was that placement of individual instances seems unfeasible and thus, some aggregation must be used. This creates the need for additional information to group data objects.

4.4.2   Available Information

This decision defines the amount of information that is available to the dynamic memory manager during an allocation. The tree in Fig. 4.3b shows several non-exclusive options that can be combined according to the needs of the design. Traditionally, the DMM knew only the size of the memory request and the starting address of the blocks being freed because that was enough for the tasks entrusted to the dynamic memory mechanism of most general-purpose languages. Typical considerations were building independent lists of blocks based on sizes to find a suitable free block quickly or the inclusion of coalescing and splitting to reduce the impact of fragmentation.

That simple interface is clearly not enough to implement data placement through the DMM because it would not have enough information to distinguish and allocate in different heaps instances of the same size but from different DDTs. Without the ability to exploit additional information, the designer has three main options: (a) Place all the dynamic objects in a single pool and rely on caching at run-time; (b) place all the dynamic objects in a single pool, measure the total number of accesses and use static data techniques to assign resources to the whole pool; and (c) analyze the amount of accesses for each allocation size and assign resources accordingly, disregarding that instances of different DDTs can receive each a quite different number of accesses even if they have a similar size.

A second possibility is that the DMM receives the size of the request and the type of the object that is being allocated (or destroyed). With that information the DMM can use the characterization of the typical behavior of the instances of the corresponding DDT, such as allocation and deallocation patterns and mean number of accesses to each instance, and assign resources from specific heaps to the instances of different DDTs. Here lies a key insight to implement data placement using the dynamic memory manager:

> The dynamic memory manager needs to know the type of the objects on which it operates to establish a correlation with the known properties of their DDTs.

However, we previously saw that this approach has a serious drawback: Heaps assigned in exclusivity to instances of a single DDT can undergo significant periods of underutilization and those resources cannot be claimed to create instances of other DDTs.

In this work we propose a third option: Providing the DMM with knowledge about the compatibilities between DDTs. With that information, the DMM can implement a controlled amount of resource sharing to increase exploitation without impacting performance—as it will be able to limit the likelihood of seldom accessed

DDTs blocking resources from more critical ones. This option enables a trade-off between exclusivity (in resource assignment) and exploitation. In compensation for the need to modify the dynamic memory API to include the additional information, the run-time implementation of this mechanism requires no additional hardware at run-time.

Finally, Fig. 4.3b shows also a fourth option: Making the source code location of the call to the dynamic memory API available to the DMM itself. Although this information is not exploited by the methodology that we explore in this text, it could be used in the future to distinguish instances of the same DDT with dissimilar behaviors. For example, nodes of a container structure used in different ways in various modules of the application. However, this information would not be enough on its own to cope with objects—such as dynamically-allocated arrays— whose placement needs to change according to the size of each instance. Location information can be easily obtained in languages such as c or C++ through macros. For example, hashing the values of the macros __FILE__, __LINE__ and __func__.

4.4.3   Manager Structure

A pool can be managed as a global monolithic entity or as a collection of individual pools (Fig. 4.3c). A monolithic pool consists of a single all-knowing dynamic memory manager handling one or more heaps of memory. The space in the heaps can be divided, for example, into lists of blocks of specific sizes; coalescing and splitting operations can move blocks between lists as their sizes change. The DMM can take more global decisions because it knows the state of all the heaps.

On the contrary, a modular pool comprises several independent dynamic memory managers, each controlling one or more heaps. The main advantage is that each DMM can be tailored to the characteristics of the DDTs assigned to it, providing optimized specific data structures and algorithms for the heaps that it controls as if no other DDTs were present in the application. The DMM that has to serve each application request can be selected according to the request properties (e.g., using the type of the object that is being allocated if DDT information is available) or through a cascaded mechanism where the DMMs are probed in turn.

Finally, modular designs that include a top layer to take global decisions are also possible. For example, a global manager that moves free blocks from the heap of the consumer thread to the heap of the producer thread in a producer–consumer scheme.

4.4.4   Order of Choice

The order in which the trees of Fig. 4.3 are considered is relevant because a choice preference is placed on the later trees by the former ones, and the most expensive options in a tree need only to be chosen if required by previous decisions.

Therefore, we propose that the most useful order in the decision trees is determining first whether the DMM has to produce the placement of dynamic data objects or not. Then, the required amount of information can be determined—if that is not affordable, the next best option can be chosen, but the result may be suboptimal. Next, the designer can opt for a monolithic or modular approach. As the duties and complexity of the DMM increase, the most likely it becomes that the modular approach will be easier and more efficient.

Finally, the rest of decisions required to produce a working dynamic memory manager can be taken, as explained in previous work such as by Atienza et al. [1, 3]: Use of coalescing and splitting, keeping independent lists for different block sizes, fixed block sizes or ability to adjust them (via splitting), order of blocks in the lists of free blocks (influences the effort needed to find a suitable block for every allocation), the policy for block selection (LIFO, FIFO, first fit, exact fit, next fit, best fit, worst fit), when to apply coalescing and/or splitting (at block request or discard), etc. Those decisions are evaluated once for a monolithic pool, or once for every DMM in a modular pool.

4.5   Putting Everything Together: Summary of the Methodology

Our proposal for placement of dynamic data objects constitutes an intermediate point between dynamic placement techniques, which can introduce an excess of data movements whose cost may not be compensated with a high access locality, and static placement of individual DDTs, which risks a low exploitation of resources. It consists on creating groups of DDTs with similar characteristics or dissimilar footprint requirements before performing the assignment of physical resources. As these processes are complex, we propose a set of heuristics to solve them efficiently; experimental data show the improvements that can be attained.

An independent dynamic memory manager controls the space assigned to each group, making up the application's pools. We further propose to use it to implement the calculated placement at run-time. This choice stems from the fact that the size and creation time of dynamic objects is unknown until the moment when they are allocated. Therefore, it seems appropriate to seize that chance to look for a memory block not only of the appropriate size, but also in the most suitable memory resource. Nevertheless, dynamic memory management can affect significantly the performance of the whole system in some applications, so the process must introduce as little overheads as possible.

Extensive profiling and analysis supplies information about the DDT character-istics and the analysis of compatibilities. Object size and type are provided to the

DMM (in the case of C++) via an augmented dynamic memory API. To this end, class declarations in the application are instrumented (as explained in Sect. 4.6) to first profile the application and, then, transparently implement the extended API. At run-time, the DMM uses this extra information to select the appropriate pool to serve each request. The selection mechanism can be implemented as a simple chain of selections favoring the common cases or, in more complex situations, as a lookup table, and should not introduce a noticeable overhead in most cases.

We propose to use a modular structure not only to ease the creation of DMMs specifically tailored for the DDTs in a pool, but also because the very structure of a modular DMM includes implicitly the knowledge about the DDTs that can be grouped together: The DDTs in a group are all processed by the DMM of the corresponding (sub)pool. In other words, two DDTs share memory resources if they are in the same group. Each (sub)DMM can look for blocks to satisfy memory requests disregarding object types because the mapping phase provided it with the most appropriate memory resources for its assigned DDTs and it will not be asked to allocate extraneous ones. Thus, the grouping information generated at design time is implicitly embedded in the DMM structure and can be exploited at run-time with a simple mechanism for (sub)pool selection.

The choice of a modular design has two additional benefits. First, free blocks in a pool cannot be used by a different DMM to satisfy any memory request, which is a desirable effect because it saves further mechanisms to reclaim resources when more important objects are created. Second, the DMM of a pool that hosts objects of several sizes may use coalescing and splitting as suitable, disregarding the block sizes served by other pools.

The following sections describe in detail each of the steps of the methodology as we envision them (Fig. 1.8).

4.6 Instrumentation and Profiling

In this section we document the instrumentation techniques specifically used for our work on data placement, which have the dual purpose of profiling the applications during the design phase and providing extended (DDT) information for the DMM at run-time. A more general approach for the characterization of the dynamic data behavior of software applications is documented by Bartzas et al. [6] as part of a framework for extraction of software metadata.

These techniques allow extracting data access information at the data type abstraction level with minimal programmer intervention. Obtaining information at the data type level is important to implement data type optimizations [22] and dynamic data object placement on memory resources. In turn, reduced programmer intervention is important, particularly during early design phases, to minimize the overhead of adapting the instrumentation to later modifications that may affect significantly the structure of the code. The impact of instrumentation is limited to:

- The inclusion of one header file (.hpp) per code file (.cpp), which can be avoided if precompiled headers are used.
- The modification of one line in the declaration of each class whose instances are created dynamically.
- Every allocation of dynamic vectors through direct calls to `new`.
- Every direct call to `malloc()`.

The instrumentation phase requires modifying those classes whose instances are created dynamically so that they inherit from a generic class that overloads the `new` and `delete` operators and parameterizes them with a univocal identifier for each DDT. Direct calls to the functions `malloc()` and `free()` need also to be modified to include such a univocal identifier. That modifications alone are enough to track the use of dynamic memory in the application. However, to profile also accesses to data objects, we have developed a mechanism based on virtual memory protection to profile data accesses avoiding further source code modifications—which would anyways be useless for the final deployment. In this section, we explain the basics of the template-based mechanism for extraction of allocation information and the use of virtual memory support to obtain data access information without additional source code modifications.

During the profiling phase, the designer has to identify the most common execution scenarios and trigger the application using representative inputs to cover them. This enables the identification of different system scenarios and the construction of a specific solution for each of them. The instrumentation generates at this stage a log file that contains an entry for every allocation and data access event during the application execution. This file, which generally has a size in the order of gigabytes (GB), is one of the inputs for our methodology.

At run-time, the same instrumentation is reused to supply the DMM with the DDT of each object allocated or destroyed. Therefore, the overhead imposed on the designers should be minimal given the significant improvements that can be attained. Deployment in environments different to C++ is possible given an equivalent method to supply DDT information to the DMM.

4.6.1   Template-Based Extraction of Allocation Information

The following fragment of code shows the class that implements the logging of allocation events:

```
template <int ID>
class allocatedExceptions {
  public:
  static void * operator new(const size_t sz) {
    return logged_malloc(ID, sz);
  }
  static void operator delete(void * p) {
    logged_free(ID, p);
  }
  static void * operator new[](const size_t sz) {
    return logged_malloc(ID, sz);
  }
  static void operator delete[](void * p) {
    logged_free(ID, p);
  }
};
```

The class `allocatedExceptions` defines class-level `new` and `delete`. During profiling, the original requests are normally forwarded (after logging) to the system allocator through the underlying `malloc()` and `free()` functions. In contrast, at run-time the requests are handled by the custom memory manager used to implement data placement. Thus, the promise of serving both purposes with the same instrumentation is fulfilled.

To instrument the application, the designer needs only to modify the declaration of the classes that correspond to the dynamic data objects of the application so that they inherit from `allocatedExceptions`, parameterizing it with a unique identifier. No other lines in the class source code need to be modified:

```
class NewClass : public allocatedExceptions<UNIQUE_ID> {
  ...
};
```

4.6.2   Virtual Memory Support for Data Access Profiling

We have developed a technique for profiling of dynamic-data accesses that trades human effort for execution performance during profiling and that does not require modifications of the operating system kernel. The main drawback of this technique is that it requires support for virtual memory in the platform. If the target platform lacks this functionality, it may still be possible to get an approximation of the overall application behavior using a different platform.

4.6.2.1 Mechanism

Our technique, which we shall refer to as "exception-based profiling," consists on creating a big heap of memory and removing access permissions to all its memory pages. An exception handler is defined to catch subsequent application accesses. To manage the space in the heap, a custom DMM is designed and used by the application through the normal instrumentation. When the application accesses a data object in the heap, the processor generates automatically an exception. The exception handler uses the information provided by the operating system in the exception frame to identify the address that was being accessed by the application (and whether it was a read or a write access). Then, it enables access to the memory page containing that address. Before instructing the operating system to retry the offending instruction in the application, the exception handler activates a special processor mode that generates an exception after a single instruction is executed. This mechanism is commonly used by debuggers to implement "single-step execution."

After the data access instruction is executed by the application, the processor generates an exception and the exception handler recovers control. The next action is to revoke again access permissions to the heap page accessed by the application. Then, the single-step mode is disabled, the access is recorded in the log file that contains the profiling information and execution of the application is resumed normally until the next access to a data object in the heap. Figure 4.4 illustrates the whole process.

Although those details are irrelevant for the purpose of profiling the number of data accesses, it may be interesting to mention that the custom designed DMM uses a container (specifically, an `std::multimap`) built outside of the heap that contains pairs <address, size> ordered by block address. This setup avoids that accesses performed by the DMM itself, which is yet to be defined, are included with the rest of accesses of the application.

4.6.2.2 Implementation

To add exception-based profiling to an application, the only modification required is to substitute its `main()` function with the one contained in the profiling library, renaming it. After initialization, the `main()` method provided by the library calls the entry point of the application. The following code fragments illustrate the implementation of the core methods in the library for the Microsoft Windows© operating system[8]:

---

[8]The source code presented through this work is for illustrative purposes only and should not be used for any other purposes. In particular, most checks for return error codes are omitted.

```
static volatile void * theHeap;
static volatile bool inException = false;
static FILE * logFile;

int main(int argc, char ** argv) {
  unsigned long foo;

  // Create the heap
  theHeap = VirtualAlloc(NULL, HEAP_SIZE, MEM_RESERVE, PAGE_NOACCESS);
  if (theHeap == NULL)
  return -1;

  // Commit individual pages
  for (unsigned long offset = 0; offset < HEAP_SIZE; offset += 4096) {
    if (VirtualAlloc((unsigned char*)theHeap + offset, 4096, MEM_COMMIT,
    PAGE_READWRITE) == NULL)
    exit(-1);
  }

  // Protect heap pages
  VirtualProtect((void *)theHeapStart, HEAP_SIZE, PAGE_NOACCESS, &foo);
  inException = false;

  // Create logging file
  logFile = CreateLogFile(PATH_TO_LOG_FILE);
  atexit(CloseProfiling);

  // Start application inside an exception handler
  __try {
    MainCode(argc, argv);
  }
  __except (ResolveException(GetExceptionCode(),
  GetExceptionInformation())) {
    // The handler always resumes execution, so nothing to do here.
  }

  return 0;
}

...

void CloseProfiling() {
  unsigned long foo;
  ...
  fclose(logFile);

  // Unprotect heap pages so it can be deleted.
  VirtualProtect((void*)theHeapStart, HEAP_SIZE, PAGE_READWRITE, &foo);
  // Free the whole heap.
  VirtualFree((void*)theHeap, 0, MEM_DECOMMIT | MEM_RELEASE);

  ...
}
```

```
...

int ResolveException(int exceptCode, EXCEPTION_POINTERS * state) {
  unsigned long foo;

  // Extract context information for the exception
  EXCEPTION_RECORD * infoExcept = state->ExceptionRecord;
  CONTEXT * infoContext = state->ContextRecord;

  if (exceptCode == EXCEPTION_ACCESS_VIOLATION) {
    // First access attempt by the application
    if (inException) // Error: double virtual exception
      exit(-1);

    // Enable access for the correct memory page
    lastAddress = (void *)infoExcept->ExceptionInformation[1];
    VirtualProtect((void*)lastAddress, 16, PAGE_READWRITE, &foo);

    infoContext->EFlags |= 0x0100; // Activate single-step flag.

    // Register operation into log file.
    // If (infoExcept->ExceptionInformation[0] == 0), it was a read.
    // Otherwise, it was a write.

    inException = true;
    // Return to the application to retry the memory access
    return EXCEPTION_CONTINUE_EXECUTION;
  }
  else if (exceptCode == EXCEPTION_SINGLE_STEP) {
    // Single-step after the application executes the access

    // Protect again the affected memory page
    VirtualProtect((void*)lastAddress, 16, PAGE_NOACCESS, &foo);

    infoContext->EFlags &= 0xFEFF; // De-activate single step flag.

    inException = false;
    return EXCEPTION_CONTINUE_EXECUTION;
    // Return to the application and continue normal execution
  }
  else { // Unhandled exception
    exit(-1);
  }
}
```

Heap space is reserved once at the beginning of the execution to ensure a continuous range of addresses. However, its pages are committed individually because VirtualProtect() works (empirically) much faster in that way.

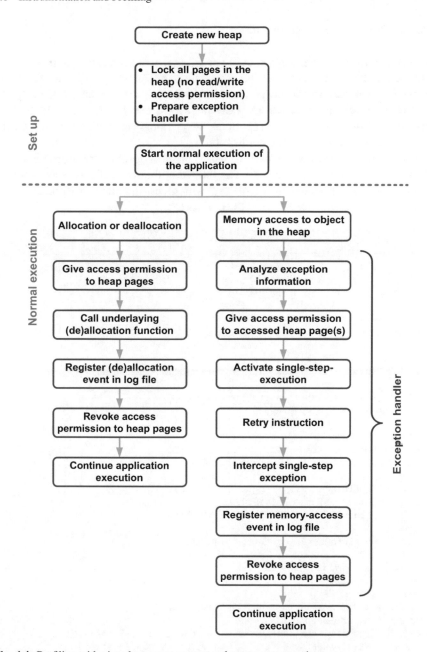

Fig. 4.4 Profiling with virtual memory support and processor exceptions

4.6.2.3 Performance Optimization

The performance of the exception-based profiling mechanism can be improved with a low-level trick. The previous implementation raises two exceptions for every access to an object in the heap. However, the designer can easily identify the most common operation codes that raise the access exceptions and emulate them explicitly in ResolveException(). In that way, a single exception is raised for the most common operations in the application.

This technique can also be used to tackle memory-to-memory instructions that might otherwise access two pages.[9] The following fragment of code shows how to emulate the opcode sequence "8B 48 xx," which corresponds in x86 mode to the instruction mov ecx, dword ptr [eax + xx]:

```
unsigned char * ip;
ip = (unsigned char *)infoContext->Eip;

// Examine the bytes at *ip
// 8B 48 xx mov ecx, dword ptr [eax + xx]

// Get the offset
signed char offset = (signed char)*(ip + 2);

// Read the value from memory
UINT32 * pValue = (UINT32 *)(infoContext->Eax + offset);

// Update the destination register
infoContext->Ecx = *pValue;

// Update eflags!
if (infoContext->Ecx == 0)
  SET_BIT(infoContext->EFlags, FLAG_ZERO);
else
  CLEAR_BIT(infoContext->EFlags, FLAG_ZERO);
if (infoContext->Ecx & 0x80000000)
  SET_BIT(infoContext->EFlags, FLAG_SIGN);
else
  CLEAR_BIT(infoContext->EFlags, FLAG_SIGN);

// Skip the emulated instruction
infoContext->Eip += 3;

// Protect the memory page and continue without
// entering the single-step mode
vRes = VirtualProtect(lastAddress, 16, PAGE_NOACCESS, &foo);
infoContext->EFlags &= 0xFEFF; // De-activate single step flag.
inException = false;
```

---

[9]Other possible solutions to this problem are identifying the corresponding opcodes and enabling accesses to both pages at the same time, or allowing nested access exceptions, granting and revoking permissions successively to both pages.

Similar optimizations may be possible to improve performance in other processor architectures, adjusting the specific opcodes used.

4.6.3 Summary

Profiling memory allocation operations can be done easily with multiple techniques and tools. However, profiling memory accesses in an exact way with no dedicated hardware support nor kernel-mode drivers and without impact on application performance is more complex. In this text we propose two different techniques for profiling memory allocations and memory accesses that can be reused to pass data type information to the DMM at run-time.

The designer has to keep in mind that profiling may alter slightly the behavior of the application. Moreover, with the exceptions-based technique, processor registers are used normally and accesses to variables (or class members) stored in them will not be recorded, which may be relevant if the initial profiling is performed on a different processor architecture than that used in the final platform. However, due to the nature of dynamic structures and the extra level of indirection used to access dynamic objects through a pointer (or reference), it is quite possible that the interference of any of these techniques is minimal. Particularly because compilers tend, in general, to optimize into registers only local (stack) variables, because they cannot be sure of the accesses performed by other functions on the same variables.

4.7 Analysis

The analysis step extracts the following information for each DDT:

- **Maximum number of instances** that are concurrently alive during the application execution.
- **Maximum footprint:** Size of data element × maximum number of instances alive.
- **Number of accesses** (reads and writes). This information is extracted for every single instance, but is later aggregated for the whole DDT because our methodology does not currently process instances individually.
- **Allocation and deallocation sequence,** counting every instance of the DDT ever created.
- **Frequency of accesses per byte (FPB)**: Number of accesses to all the instances of a DDT divided by its maximum footprint.
- **"Liveness:"** Evolution of the DDT footprint along execution time as a list of footprint variations.

The analysis tool can distinguish between instances of the same DDT created with different sizes (e.g., multiple instances of a dynamically-allocated array created

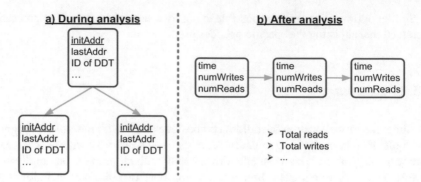

Fig. 4.5 Data structures during the analysis step: (**a**) Ordered tree with blocks active up to the current profiling event. (**b**) Behavior for each DDT. In this context, "time" refers to the number of allocation events since the beginning of the execution

at the same location in the source code). In the rest of this text this is used as a specialization of the DDT concept, that is, different sized instances are considered as distinct DDTs.

The analysis algorithm is straightforward: Each packet in the log file produced during profiling is analyzed. For every allocation event, an object representing the allocated block is created and introduced in a tree (`std::map<>`) ordered by allocation address (Fig. 4.5a)—that is, the address assigned in the original application. Then, for every memory access event the algorithm looks in the tree for the block that covers the accessed address and updates its access counters. Finally, when a deallocation event is found, the object representation is extracted from the tree and destroyed. This mechanism allows identifying the instance (and its DDT) that corresponds to each memory access recorded in the log file, tracking the addresses that were assigned to each one during the execution of the original application.

The output of the analysis step (Fig. 4.5b) is a list of allocation events for each DDT, where each event includes the number of read and write accesses to all the DDT's alive instances since the previous event for that DDT. This list is deemed as the DDT behavior. Liveness information, which can be extracted directly from the allocation events in the DDT behavior, is used during the grouping step to identify DDTs that can share the same pool.

The use of an independent list of events for each DDT, instead of a single list for all the DDTs in the application with empty entries for the DDTs that do not have footprint variations on a given instant, reduces the memory requirements of the analysis tool itself and improves its access locality, which is an interesting issue on its own to reduce the time required to complete or evaluate a design.[10]

---

[10]Interestingly, some details of the internal implementation of our tool profited from our profiling and analysis techniques themselves.

4.8   Group Creation

As we have seen previously, placement of dynamic data objects over a heterogeneous memory subsystem is a hard problem. The very dynamic nature of the objects
makes foreseeing techniques to produce an exact placement highly unlikely. For that
reason, we have proposed a methodology based on three concepts: First, performing
placement at the DDT level. Second, analyzing the properties of the DDTs to group
those with similar characteristics and place them indistinctly. Third, mapping the
resulting groups into memory resources according to their access characteristics.

Grouping is the central idea in this approach. It selects a point between assigning
each DDT to a separate group (optimal placement, worst exploitation of resources)
and assigning all of them to a single group (no specific placement, best exploitation
of resources). DDTs assigned to different groups are kept in separate pools and, if
possible, they will be placed in different memory resources during the mapping step.
Similarly, DDTs with complementary characteristics that are included in the same
group will be managed in the same pool; thus, their instances will be placed on the
resources assigned to the pool indistinctly.

Grouping has two main tasks: First, identifying DDTs whose instances have
similar access frequencies and patterns, so that any of them will benefit similarly
of the assigned resources. Second, balancing between leaving resources underused
when there are not enough alive instances of the DDTs assigned to the corresponding group, and allowing instances from less accessed DDTs to use better memory
resources when there is some space left in them. The grouping algorithm approaches
this second task analyzing DDTs that have complementary footprint demands along
time, so that valleys in the footprint of some DDTs compensate for peaks of the
others and the overall exploitation ratio of the memory resources is kept as high as
possible during the whole execution time.

To reduce the complexity of grouping, the approach that we explore here is
based on a greedy algorithm that offers several "knobs" that the designer can use
to steer the process and adapt it to the specific features of the application under
development. This solution is not optimal, yet in Chap. 6 we will see that it already
attains significant performance advantages for several case studies.

4.8.1   Liveness and Exploitation Ratio

The following two metrics are relevant to minimize the periods when memory
resources are underused:

- **Group liveness:** Similarly to the case of individual DDTs, the liveness of a group
  is the evolution along time of the combined footprint of all the DDTs that it
  contains. It can be implemented with a list of memory allocation events that
  represents the group "behavior."

- **Exploitation ratio:** The occupation degree of a group (or pool) along several time instants:

$$Exploitation\ ratio = \frac{\sum_{t=1}^{N} \frac{Required\ footprint(t)}{Pool\ size}}{N}$$

In essence, the exploitation ratio provides a measure of how well the grouping step manages to combine DDTs that fill the space of each memory resource during all the execution time. During the grouping step, as the size of each group is not yet fixed, it is calculated as the occupation at each instant respect the maximum footprint required by the DDTs already included in the group. In this way, the grouping algorithm tries to identify the DDTs whose liveness is complementary along time to reduce their combined footprint, but it may also add DDTs with a lower FPB to a group if the exploitation ratio improves and the total size does not increase (more than a predefined parameter)—the "filling the valleys" part. During simulation, the exploitation ratio can be calculated for the final pools (which do have a fixed size) to check the effectiveness of a solution.

These concepts can be illustrated using a hypothetical application with two threads as an example. The first thread processes input events as they are received, using DDT_1 as a buffer. The second one consumes the instances of DDT_1 and builds an internal representation in DDT_2, reducing the footprint of DDT_1. Accordingly, the footprint of DDT_2 is reduced when the events expire and the related objects are deleted. Figure 4.6a shows the liveness of each DDT. The maximum footprints of the DDTs are 7 KB and 10 KB, respectively. Therefore, if both DDTs were placed independently (that is, in two different pools), the total required footprint would be 17 KB (labeled as "Added" in the figure). However, grouping both DDTs together reduces the maximum required footprint to 11 KB (labeled as

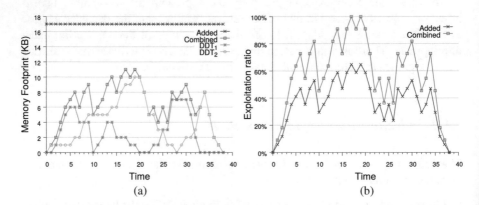

Fig. 4.6 Liveness and exploitation ratio for two hypothetical DDTs. (**a**) Liveness (evolution of footprint) for the two DDTs considered independently, their added maximum footprints and the footprint of a group combining them. (**b**) Exploitation ratio for the case of an independent group for each DDT (38.2% on average) and for one combined group (59.1% on average)

"Combined"). Figure 4.6b compares the exploitation ratio of the group with the exploitation ratio that would result if the two DDTs were kept apart. In this hypothetical case, grouping would increase the exploitation ratio of the memory resources, thus reducing the required footprint. These two DDTs can be kept apart from the rest of DDTs of the application, or they might be further combined with others.

4.8.2   Algorithm Parameters

The designer can use the following parameters to guide the grouping algorithm:

1. **MaxGroups.** The maximum number of groups that the algorithm can create. This parameter controls the trade-off between minimizing memory footprint (for a value of 1) and creating multiple pools (for values bigger than 1). It should be at least as big as the number of memory modules of distinct characteristics present in the platform; otherwise, the chance to separate DDTs with different behaviors into independent memory modules would be lost.

   In comparison with the bin packing problem, the grouping algorithm does not try to fit all the DDTs in the maximum allowed number of groups; instead, DDTs that are not suitable for inclusion in any group are simply pushed into the last group, which contains the DDTs that will probably be placed on main memory.

2. **MaxIncMF$_G$.** The maximum ratio that the footprint of group G is allowed to increase when a DDT is added to that group. This parameter introduces some flexibility while combining DDTs, which is useful because their footprints will usually not be a perfect match.

3. **MinIncFPB.** Minimum ratio between the old and the new FPB of a group that needs to be achieved before a DDT is added to that group. The default value of 1.0 allows any DDT that increases the FPB of the group to be included.

4. **MinIncExpRatio.** The minimum increase in the exploitation ratio that allows a DDT with a lower FPB to be included in a group. The default value of 1.0 allows any DDT that increases the exploitation ratio of the group to be included in it.

5. **ExcludedDDTs.** If the designers have good knowledge of the application, they can decide to exclude a DDT from the grouping process and manually push it into the last group for placement on main memory.

The combination of MaxIncMF, MinIncFPB, and MinIncExpRatio allows balancing between increasing the exploitation ratio of the pools, maximizing the FPB of their DDTs and bounding the number of distinct DDTs that can be added to a group. As an example, a high MaxIncMF and a MinIncFPB of 1.0 will favor grouping many DDTs together. Increasing the value of MinIncFPB will prevent the addition of DDTs with very few accesses, even if they apparently match the valleys in the footprint of the group (this can be useful, for example, if the designers know that some DDT has a very dynamic behavior that is difficult to capture during profiling).

4.8.3 Algorithm

Algorithm 1 presents the pseudo-code for the grouping process. The main idea is building groups until all the DDTs have been included in one or the maximum number of groups is reached. Each time the algorithm builds a new group, it checks all the remaining DDTs in order of descending FPBs. If the DDT is compatible with the current contents of the group, that is, its peak requirements match the footprint

---

Algorithm 1 Grouping

 1: **function** GROUPING(DDTs : List of DDTs) : List of Groups
 2:     Order the DDTs on descending FPB
 3:     Exclude the DDTs that were marked by the designer (ExcludedDDTs)
 4:     **While** there are any DDTs remaining **and** MaxGroups is not reached **do**
 5:         Create a new group
 6:         **For** each remaining DDT **do**
 7:             Calculate the liveness and FPB that would result if the DDT were
                 added to the group (CALCNEWFOOTPRINTANDFPB)
 8:             **If** the new behavior passes the tests in CHECKCOMPATIBILITY **then**
 9:                 Add the DDT to the group
10:                 Remove the DDT from the list of DDTs
11:     Push any remaining DDTs into the last group
12:     Add the DDTs that were excluded to the last group
13:     Order the groups on descending FPB
14:     **Return** the list of groups
15: **end function**

16: **function** CHECKCOMPATIBILITY(newBehavior : Behavior) : Boolean
17:     **Return** true **if** the new footprint does not exceed the maximum footprint for any group
18:         **and** the footprint is not incremented more than MaxIncMF$_G$
19:         **and** the FPB of the group is increased by at least MinIncFPB
20:         **and** the exploitation ratio is increased by at least MinIncExpRatio
21: **end function**

22: **function** CALCNEWFOOTPRINTANDFPB(Group, DDT) : Behavior
23:     Create a new behavior
24:     **While** there are events left in the behavior of the group or the DDT **do**
25:         Select the next event from the group and the DDT
26:         **If** both correspond to the same time instant **then**
27:             Create a new event with the sum of the footprint of the group and the DDT
28:         **Else if** the event of the group comes earlier **then**
29:             Create a new event with the addition of the current footprint of the group
                 and the last known footprint of the DDT
30:         **Else** {(} if the event of the DDT comes earlier)
31:             Create a new event with the addition of the current footprint of the DDT
                 and the last known footprint of the group
32:         Update the FPB with the maximum footprint registered and the sum of reads and writes
             to the group and the DDT
33:     **Return** new behavior
34: **end function**

---

minima of the group, then it is included. Otherwise, it is discarded and the next one is considered. This increases the exploitation ratio of the group as more instances will be created using the same amount of resources. The available parameters allow relaxing the restrictions applied for joining DDTs to existing groups so that, for instance, a small increase on the group footprint is allowed for DDTs with similar FPBs that do not have a perfectly matching liveness.[11]

The DDTs are evaluated in order of decreasing FPB to ensure that if a DDT matches the liveness of a group, it is the DDT with the highest FPB among the remaining ones—thus, observing the heuristic of placing first the DDTs with the highest density of accesses per size unit. Once the FPB of the resulting group is known, a check is made to verify that a minimum increment is achieved. This is useful, for instance, to avoid including DDTs with low FPBs that could hinder the access pattern of the group once it is placed into memory resources such as DRAMs.

The group liveness is kept updated as a combination of the liveness of the DDTs in it. This ensures that no comparisons between individual DDTs, but between the DDTs and the groups, are executed, reducing the algorithm complexity. To evaluate the inclusion of a DDT in a group, the new combined behavior is calculated (lines 22–34). This is a straightforward process that involves combining two ordered lists (the lists of allocation events of the group and the new DDT) and accounting for the accumulated footprint and accesses of the group and the DDT. Then, the constraints imposed by the grouping parameters are evaluated for the new behavior (lines 16–21). New criteria such as the amount of consecutive accesses or specific access patterns can be easily incorporated into this check in the future.

The output of this step is a list of groups with the DDTs included in each of them and their aggregated characteristics (maximum footprint, liveness, FPB and exploitation ratio). The worst-case computational complexity of the algorithm is bounded by $O(n^2m + nm)$, where n is the number of DDTs in the application and m is the number of entries in the liveness of the DDTs. However, as normally n is in the order of tens of DDTs while m is in the order of millions of allocation events per DDT, $m \gg n$ and the complexity is in practical terms closer to $O(m)$.

4.8.3.1 Justification

The grouping algorithm presents two extreme cases. In the first one, it creates as many groups as DDTs in the application. In the second, all the DDTs require memory mostly at disjoint times and can be combined in a single pool. In the first case, the algorithm performs $O(n^2)$ tests, each test requiring $O(m + m)$ operations to produce the behavior of the new (hypothetical) group—merging two sorted lists

---

[11]This is one example of the difficulty of the grouping problem: What is a better option, to combine those two DDTs with similar FPBs at the expense of increasing the size of the group (i.e., requiring more resources during mapping), or leave them apart in case that other DDT with a better footprint match with the group may be found later?

with v and w elements respectively has a complexity of $O(v + w)$. Thus results a complexity of $O(n^2m)$.

However, in the second case the algorithm performs $O(n)$ tests (one test for each DDT against the only extant group). Each of those tests requires $O(nm + m)$ operations to generate the new behavior—the m events in the liveness list of the n DDTs accumulate in the liveness of the group, which is tested against the m elements in each newly considered DDT. Since the time of the tests dominates the time of sorting, the worst-case complexity is $O(n(nm + m)) \equiv O(n^2m + nm)$.

4.9   Definition of Pool Structure and Algorithms

The concept of pool represents in this methodology one or several address ranges (heaps) reserved for the allocation of DDTs and the data structures and algorithms needed to manage them. For every group from the previous step, a pool that inherits its list of DDTs is generated. During this step, considerations like the degree of block coalescing and splitting, choice of fit algorithms, internal and external fragmentation, and number of accesses to internal data structures and their memory overhead take place. The result of this step is a list of pools ordered by their FPBs and the description of the chosen algorithms in a form of metadata that can be used to build the memory managers at run-time.

The construction of efficient DMMs is a complex problem that, as outlined in Sect. 2.3, has been the subject of decades of study. Among other resources widely available in the literature, the Lea allocator used in Linux systems is described by Lea [19], an extensive description of a methodology that can be used to implement efficient custom dynamic memory management is presented by Atienza et al. [3, 4] and a notable technique to efficiently manage a heap that has been placed on a scratchpad memory is presented by McIlroy et al. [20]. Therefore, this process is not further described here.

Similarly, efficient composition of modular dynamic memory managers can be achieved with techniques such as the mixins used by Berger et al. [7] and Atienza et al. [2, 3] or through call inlining as used by Grunwald and Zorn in CustoMalloc [15]—although the latter can only be used for objects whose size is known by the compiler, that is precisely the case for common constructs such as struct (record) types.

As an example of the type of decisions involved during the construction of an efficient DMM, consider the case of a pool that receives allocation requests of 13 B and 23 B. The designer of the DMM can take several options. For instance, the DMM can have two lists of free blocks, one for each request size. If the heap space is split between the two sizes, then the DMM does not need to add any space for the size of each block: The address range of the block identifies its size. However, in that case the DMM will not be able to reuse blocks of one size for requests of the other— in line with the idea that the DDTs in one pool can share the space assigned to it. To implement coalescing and splitting, the DMM will need to add a size field to each memory block. Assuming it uses 4 B for this purpose, the size of the blocks will be 17 B and 27 B, with an overhead of 31% and 17%, respectively. To complicate things further, coalescing two smaller blocks will create a block of 34 B that is slightly big for any request. Depending on the characteristics of the application and the success of the grouping process in finding DDTs with complementary liveness, the DMM may be able to coalesce many blocks of one size at some point during execution, recovering most of the space as a big area suitable for any number of new requests.

After such considerations, one could be tempted to execute the pool construction step before grouping, so that the DDTs in a pool are chosen to ease the implementation of the DMM. However, such decision would defeat the whole purpose of placement because DDTs with many accesses might be joined with seldom accessed ones, creating groups with lower overall FPB. Although this could be the topic of future research, in general it seems preferable to sacrifice some space to obtain better access performance—especially because the grouping step is designed with the specific purpose of finding DDTs with complementary liveness.

In summary, the numerous and complex considerations during the construction of the DMM are out of the scope of this work. Therefore, in this work we treat this step as a stub and pass the groups directly to the mapping step. The experiments presented in Chap. 6 using our simulator use simple ideal DMMs that help to compare the relative quality of different placement options.

4.10 Mapping into Memory Resources

The mapping step produces a placement of the pools into the available memory resources. The input to this step is the ordered list of pools with their internal characteristics (e.g., size, FPB) and a description of the memory subsystem of the platform. This description is very flexible and allows specifying different types of organizations based on buses and memory elements. The result of this step is a list of pools, where each pool is annotated with a list of address ranges that represent its placement into memory resources. The computational complexity of the mapping algorithm is in the order of $O(n)$, being n the number of pools—and assuming that the number of pools is higher or in the order of the number of memory resources.

The design of the mapping algorithm makes some assumptions. First, that pools can be split over several memory resources even if they are not mapped on consecutive addresses. This is technically feasible with modern DMMs at the expense of perhaps a slightly larger fragmentation—however, if the pools are mapped into memory blocks with consecutive memory addresses, this overhead disappears because the blocks become a single entity at the logical level. With this consideration, the mapping part of the dynamic data placement problem can be seen as an instance of the fractional (or continuous) knapsack problem [8], which can be solved optimally with a greedy algorithm. This is an important result because the mapping step could be moved in the future to a run-time phase, allowing the system to adapt to different implementations of the same architecture or to achieve a graceful degradation of performance as the system ages and some components start to fail. It could even be useful to adapt the system to changing conditions, such as powering down some of the memory elements when energy is scarce (e.g., for solar-powered devices on cloudy days).

Second, the mapping step assumes that only one memory module can be accessed during one clock cycle. It is possible to imagine situations where the processor has simultaneous access to several independent buses or where at least slow operations can be pipelined, which may be useful for memories that require several cycles per access (DRAMs fall in this category for random accesses, but they can usually transfer data continuously in burst mode). However, to efficiently exploit those extra capabilities a more complex approach would be needed. In this regard, an interesting study, limited to static data objects, was presented by Soto et al. [24]. This may constitute an interesting topic for future research.

The third assumption is that all the instances created in the pool have the same FPB. As explained in Sect. 4.1.3, discriminating between instances with a different number of accesses seems to be a complex problem, especially for short-lived instances created and destroyed in big numbers. For longer-lived ones, future research could consider hardware or compiler-assisted methods to identify very accessed instances (although such techniques would likely introduce unacceptable costs), and techniques similar to the ones used by garbage collectors to migrate them to the appropriate memories.[12]

Finally, the mapping algorithm does not take into account possible incompatibilities between pools. Such situations might arise, for example, when instances of DDTs assigned to different pools are accessed simultaneously during vector reduction operations. If the pools were both assigned to a DRAM, it would be preferable to place each of them in a different bank to reduce the number of row misses. Although the current implementation of our tools does not support this explicitly, the parameter SpreadPoolsInDRAMBanks can be used by the designer to manually implement this functionality in simple cases—indeed, such is the case in the experiments presented in Chap. 6.

4.10.1 Algorithm Parameters

The mapping algorithm offers several parameters that the designer can use to tune the solution to the application:

1. **MinMemoryLeftOver.** Minimum remaining size for any memory block after placing a pool in it. If the remaining space is lower than this value, it is assigned to the previous pool. This parameter, which prevents dealing with tiny memory

---

[12]Generational pools can be used to identify long-lived objects. However, for data placement the interest is on identifying very accessed objects among those that are long-lived.

blocks, should be set to the minimum acceptable heap size for any pool (the default is 8 B, but imposing a bigger limit may be reasonable).

2. **Backup pool.** This parameter activates the creation of the backup pool. Accesses to objects in this pool will typically incur a penalty, as the backup pool is usually located in an external DRAM. Hence, although the backup pool adds robustness to the system, its presence should not be used as an excuse to avoid a detailed profiling.

3. **PoolSizeIncreaseFactor.** The designer can adjust the amount of memory assigned to each pool with respect to the maximum footprint calculated through the analysis of group liveness. This allows balancing between covering the worst case and reducing the amount of space wasted during the most common cases. However, the experiments show that usually the liveness analysis packs enough DDTs in each group to keep a high exploitation ratio; thus, a value of 1.0 gives normally good results. The designer may still use it to explore different conditions and what-ifs.

4. **PoolSizeIncreaseFactorForSplitPools.** When a pool is split over several memory blocks (with non-contiguous address ranges), it may be necessary to increase its size to overcome the possibly higher fragmentation. The default value is 1.0, but a value as high as 1.3 was needed in some preliminary experiments, depending on the size of the DDT instances.

5. **MappingGoal.** Memory modules can be ordered according either to access energy consumption or latency. Although the case studies in Chap. 6 use memory technologies that improve both simultaneously (increasing the area per bit cost), either goal may be specified explicitly.

6. **SpreadPoolsInDRAMBanks.** This parameter spreads the pools placed in the last DRAM module (including the backup pool, if present) over its banks according to FPB ordering. The idea is to increase the probability that accesses to DDT instances from different pools hit different DRAM banks, thus reducing the number of row misses (i.e., row activations). If there are more pools than DRAM banks, the remaining ones are all placed in the last bank—of course, better options that take into account the interactions between accesses to the pools could be explored in the future.

4.10.2 Algorithm

Algorithm 2 presents the pseudo-code for the mapping algorithm. In essence, it traverses the list of pools placing them in the available memory resources, which are ordered by their efficiency. After the placement of each pool, the remaining size in the used memory resources is adjusted; memory resources are removed from the list as their capacity is exhausted.

The mapping algorithm starts by ordering the memory resources according to the target cost function, be it energy or latency. The list of pools remains ordered since the grouping step. For every pool, the algorithm tries to map as much of it as possible

---

Algorithm 2 Mapping

1: **function** MAPPING(List of pools, List of memory blocks, Cost function)
2:    Order memory blocks using the cost function (energy / access time)
3:    **For** each pool in the list of pools **do**
4:        Multiply pool size by PoolSizeIncreaseFactor
5:        **While** the pool is not fully mapped **do**
6:            Select the first block in the list of available blocks
7:            **If** the remaining size of the pool > available space in the block **then**
8:                Assign the available space in the block to the pool
9:                Remove the block from the list of memory blocks
10:               **If** the pool has only one DDT **then**
11:                   Round up pool's remaining size to a multiple of the DDT instance size
12:               **Else**
13:                   Increase the pool's remaining size by PoolSizeIncreaseFactorForSplitPools
14:           **Else**
15:               Assign the rest of the pool to the block
16:               **If** block is DRAM **and** SpreadPoolsInDRAMBanks **then**
17:                   Assign the whole DRAM bank
18:               **Else**
19:                   Reduce available space in the block
20:                   **If** available space in the block < MinMemoryLeftOver **then**
21:                       Assign everything to the pool
22:                       Remove block from the list of memory blocks
23:    **return** ($blocks$, $pools$)
24: **end function**

---

into the first memory resource in the list. If the pool is bigger than the space left in the memory resource (lines 7–13), it is split and the remaining size goes for the next round of placement. However, splitting a pool can introduce some fragmentation. Therefore, the size of the remaining portion is rounded up to the size of an instance if the pool contains only one DDT; otherwise, the designer can specify a small size increase with the parameter PoolSizeIncreaseFactorForSplitPools (lines 10–13).[13]

On the contrary, if the remaining portion of a pool fits into the current memory resource, the placement of that pool is concluded; the remaining capacity of the resource is adjusted appropriately (lines 15–22). When a pool is placed on a DRAM, the parameter SpreadPoolsInDRAMBanks is checked; if active, the whole DRAM bank is assigned to the pool in exclusivity, regardless of its actual size. In the absence of a more elaborate mechanism, the designer can use this parameter to reduce the number of row misses when the DRAM has extra capacity. For the rest of memory resources (or if the parameter is not active) the algorithm makes a final check to avoid leaving memory blocks with very small sizes (line 20).

---

[13] In platforms with many small memory resources, this adjustment can lead to an unbounded increase of the pool size. In such cases, either a smaller value should be used or the algorithm could be changed to increase the size only after the first time that the pool is split.

Finally, the mapping algorithm can produce an additional "backup pool," which will accommodate all the instances that, due to differences in the actual behavior of the application in respect to the characterization obtained during the profiling phase or to excessive pool fragmentation, cannot be fit into their corresponding pools. If present, the backup pool uses all the remaining space in the last memory resource, usually a DRAM. When the mapping step finishes, each pool has been annotated with a set of tuples in the form (memoryResourceID, startAddress, size) that represent the position in the platform's address map of each fragment of the pool.

4.10.3 Platform Description

Platform descriptions for mapping and simulation are provided through text files. As can be seen in the following examples, the syntax is very simple. It has one entry (in square brackets "[]") for every element in the memory subsystem. Valid memory types are "SRAM," "SDRAM" and "LPDDR2S2_SDRAM." Both DRAMs and SRAMs, that is, directly addressable memories, bear the label "Memory" and a unique identifier. The designer must define for them the attributes "PlatformAddress" and "Size," which define their position in the platform's address space, and "ConnectedTo," which defines their connection point in the memory subsystem. Each memory module has to define also a set of working parameters such as voltage and timing that will be used during mapping (to order memory modules according to the target cost function) and simulation.

DRAMs require several additional parameters. For example, "NumBanks" (the number of internal memory banks) and "WordsPerRow" (the number of words in the rows of each bank) control the internal configuration of the memory, while "CPUToDRAMFreqFactor," which defines the memory-bus-to-CPU frequency ratio, and "CPUFreq," which defines the CPU frequency in Hz, are used to find the memory cycle time:

$$MemCycleTime = \frac{1}{CPUFreq \,/\, CPUToDRAMFreqFactor}$$

Cache memories use instead the label "Cache." Valid cache types are "Direct" and "Associative," with "NumSets" defining the degree of associativity. Valid replacement policies are "LRU" and "Random." Our tools supports up to four cache levels per memory and multiple cache hierarchies. However, a minor limitation is currently that a cache hierarchy can be linked only to one memory module. Therefore, multiple memory modules require each their own cache hierarchy. Although this should not constitute a major applicability obstacle—most platforms contain a single DRAM module—modifying it should not represent a big challenge, either. Each cache memory defines its "Size," the length of its lines in memory words ("WordsPerLine"), its master memory ("ConnectedToBlockID") and its level in the local hierarchy ("Level"). The range of memory addresses covered by each cache memory is defined through the attributes "CachedRangeStart" and "CachedRangeSize;" using them, the designer can easily define uncached address ranges as well.

The current implementation of our methodology assumes that SRAMs are always internal (on-chip) and never sport caches. However, although not a common practice in the design of embedded platforms nowadays, it would be very easy to consider the case of external (off-chip) SRAMs used as main memories with (optional) internal caches. Similarly, on-chip DRAMs (i.e., eDRAM) and other memory technologies could also be included if needed in the future.

Finally, interconnections use the label "Interconnection," also with a unique identifier. Bus hierarchies can be defined through "RootConnection," with a value of 0 signifying that the bus is connected directly to the processor. The following fragment defines a platform with a 128 MB SDRAM and a 32 KB cache with associativity of 16 ways and lines of 4 words:

```
####################
[Memory=0]
# DRAM 128 MB
# Micron Mobile SDRAM 1 Gb
# (128 MB) MT48H32M32LF
# -6, 166 MHz, CL=3
Type="SDRAM"
ConnectedTo=1 # Interconnection ID
PlatformAddress=2147483648
Size=134217728

CPUToDRAMFreqFactor=(double)8.0
CPUFreq=(double)1332e6
NumBanks=4
WordsPerRow=1024

CL=3
tRP=3
tRCD=3
tWR=3
tCDL=1
tRC=10

vDD=(double)1.8
vDDq=(double)1.8
iDD0=(double)67.6e-3
iDD1=(double)90e-3
iDD2=(double)15e-3
iDD3=(double)18e-3
iDD4=(double)130e-3
iDD5=(double)100e-3
iDD6=(double)15e-3
cLOAD=(double)20e-12

####################
```

```
[Cache=0]
# L1-Data 32 KB
Type="Associative"
ReplacementPolicy="LRU"
ConnectedToBlockID=0
Level=0 # Caches must be
        kept ordered
# Address range cached by
        this memory:
CachedRangeStart=2147483648
CachedRangeSize=134217728
NumSets=16
Size=32768
WordsPerLine=4

EnergyRead=
(double)0.02391593739800e-9
EnergyWrite=
(double)0.02391593739800e-9
EnergyReadMiss=
(double)0.02391593739800e-9
EnergyWriteMiss=
(double)0.02391593739800e-9
DelayRead=1
DelayWrite=1  # All words in
        parallel
DelayReadMiss=1
DelayWriteMiss=1

####################
[Interconnection=1]
AcquireLatency=0
TransferLatency=0
RootConnection=0 # Processor
ConcurrentTransactions=0
TransferEnergy=(double)0.0
```

The text fragment below defines a platform with a low-power 256 MB DDR2-SDRAM and a 32 KB SRAM:

```
####################              tDQSCK_SQ=1
[Memory=0]                        tDQSS=1
# SRAM 32KB                       tCCD=1
Type="SRAM"                       tRTP=3
ConnectedTo=1 # Interconnection ID tRAS=14
PlatformAddress=0                 tRPpb=6
Size=32768                        tWTR=3
                                  tWR=5
EnergyRead=(double)               vDD1=(double)1.8
        0.00492428055021e-9       vDD2=(double)1.2
EnergyWrite=(double)              vDDca=(double)1.2
        0.00492428055021e-9       vDDq=(double)1.2
DelayRead=1                       iDDO1=(double)0.02
DelayWrite=1                      iDDO2=(double)0.047
                                  iDDOin=(double)0.006
####################              iDD3N1=(double)1.2e-3
[Memory=1]                        iDD3N2=(double)23e-3
# LPDDR2-S2 SDRAM 256MB           iDD3Nin=(double)6e-3
# (64Mx32) at 333MHz (-3)         iDD4R1=(double)0.005
Type="LPDDR2S2_SDRAM"             iDD4R2=(double)0.2
ConnectedTo=1 # Interconnection ID iDD4Rin=(double)0.006
PlatformAddress=2147483648        iDD4Rq=(double)0.006
Size=268435456                    iDD4W1=(double)0.01
                                  iDD4W2=(double)0.175
CPUToDRAMFreqFactor=(double)4.0   iDD4Win=(double)0.028
CPUFreq=(double)1332e6            cLOAD=(double)20e-12
NumBanks=8
WordsPerRow=2048                  ####################
MaxBurstLength=4                  [Interconnection=1]
DDR2Subtype="S2"                  AcquireLatency=0
                                  TransferLatency=0
tRCD=6                            RootConnection=0 # Processor
tRL=5                             ConcurrentTransactions-0
tWL=2                             TransferEnergy=(double)0.0
```

4.11 Deployment

Application deployment can be accomplished in two ways. The first one derives directly from the work previously done on DMM composition with mixins by Berger et al. [7] and Atienza et al. [2–4]. Using a library of modular components, a DMM can be built for each pool using the characteristics determined during the pool construction step. Mixins lie on one end of the trade-off between efficiency and flexibility: Although they allow for a good degree of optimization, each DMM is assembled during compilation; hence, its design is immutable at run-time—*parameters*, such as pool position (memory address) or size, can still be configured.

The second option is to define the characteristics of the DMMs and provide a set of assemblable components to construct them when the application is loaded at run-time. A combination of the factory and strategy design patterns [14] may be used to provide a set of components (strategy implementations) that can be assembled using a recipe by a factory when each pool is created. The code for the library of strategy implementations and the factory object can be shared by all the applications in the system as a dynamically linked library or shared object. This approach lies at

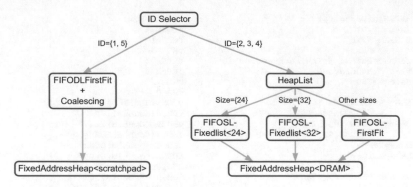

Fig. 4.7 Structure of the DMM in Example 4.1. DDTs with IDs 1 and 5 are placed in the scratchpad with a coalescing heap. DDTs with IDs 2, 3 or 4 are placed in the DRAM, with blocks of sizes 24 B and 32 B each apart from the rest

the opposite side of the trade-off, offering adaptability to changing conditions at a possibly higher cost—that future research can quantify—due to the overhead of the indirect calls introduced by the strategy pattern. Additionally, the size of the code required for the components should be lower than the sum of the code of several full DMMs.

The final deployment of the application comprises its compiled code, including the mixins-based library of DMMs, or, alternatively, its compiled code, metadata with the description of the pools, the library of DMM components and the code for the factory object to assemble them. Although the application source code has to be modified to use the new memory managers—they need to receive the identifier of the DDT to which the new instances belong—these modifications are the same introduced during the profiling phase and thus, no additional effort is required.

Although we will not cover this step—the experiments in Chap. 6 use the integrated simulator—the feasibility of deployment should be sufficiently proven by the numerous times when the library-based approach has been used.

Example 4.1 Example of deployment with the mixin-based library.

Just for the sake of this example, let us assume that we have an application with five DDTs separated in two pools, as illustrated in Fig. 4.7. The first one contains the DDTs with identifiers 1 and 5, is assigned to a scratchpad and uses coalescing to make good use of the space. The second pool contains the DDTs with identifiers 2, 3 and 4, with instances of size 24 B, 32 B and several more, is placed on the main DRAM and does not use coalescing to reduce the number of non-sequential accesses.

The following simplified sketch of code shows how the first pool could be defined using the mixins library:

```
typedef
CoalesceHeap<
  HeapList<
    FIFODLFirstFitHeap, FixedAddressHeap<SCRATCH_ADDR, SCRATCH_SIZE>
  >,
  MIN_BLOCK_SIZE, MAX_BLOCK_SIZE
>
PoolA;
```

PoolA employs a coalesceable heap that uses space from either a list of free blocks or from unused space in the scratchpad. The list of free blocks uses the first-fit mechanism to locate a block for a new request. If the list does not contain a big enough block, new space is allocated from the unused part in the scratchpad. The blocks in the list are doubly linked (next and previous) to ease the process of extracting a block from the list during coalescing. Splitting happens at allocation time, as long as the remaining block is bigger than MIN_BLOCK_SIZE bytes. Similarly, coalescing happens at deallocation time, as long as the resulting block is smaller than MAX_BLOCK_SIZE bytes. HeapList<Heap1, Heap2> is a simple mechanism that asks Heap1 to handle the request; if Heap1 refuses it, then it tries with Heap2.

The second pool can be defined with the following simplified code:

```
typedef
HeapList<
  SelectorHeap<FIFOSLFixedlistHeap, SizeSelector<24> >,
  HeapList<
    SelectorHeap<FIFOSLFixedlistHeap, SizeSelector<32> >,
    HeapList<
      FIFOSLFirstFitHeap, % SL because no coalescing
      FixedAddressHeap<DRAM_ADDR, DRAM_SIZE>
    >
  >
>
PoolB;
```

Here, the pool uses first two lists of free blocks for the most common allocated sizes. The lists are singly linked to reduce memory overhead because they are not involved in coalescing operations. As each of them contains blocks of exactly one size, free blocks do not contain any header to record their size. Therefore, the overhead is reduced to one word, to store the pointer to the next block when they are in the list, or to store their size (so that it can be determined when they are released) when they are in use by the application. The pool has a last heap organized as a list of blocks that uses first-fit to find a suitable block for the remaining object sizes. Finally, extra space is obtained from the assigned space in the DRAM when needed.

The global DMM of the application can be composed as follows:

```
typedef
HeapList<
  SelectorHeap<
    PoolA,
    OrSelector<IDSelector<1>, IDSelector<5> >
  >,
  SelectorHeap<
    PoolB,
    OrSelector<
      IDSelector<2>,
      OrSelector<IDSelector<3>, IDSelector<4> >
    >
  >
>
GlobalDMM;
```

In essence, the global DMM uses the ID corresponding to the DDT of the object to find the pool that must attend the petition. `IDSelector<ID>` returns `true` if the DDT of the allocated object coincides with its parameter. `OrSelector<A, B>` returns `A || B`. Finally, `SelectorHeap<pool, condition>` attends a memory request if its condition test returns `true`; otherwise, it rejects the request.

The parameters used to configure the different heaps are passed as template arguments. Therefore, the compiler knows them and can apply optimizations such as aggressive inlining to avoid most function calls. For example, a long chain of `OrSelector` objects can be reduced to a simple `OR` expression.

References

1. Atienza, D.: Metodología multinivel de refinamiento del subsistema de memoria dinámica para los sistemas empotrados multimedia de altas prestaciones. Ph.D. thesis, Universidad Complutense de Madrid, Departamento de Arquitectura de Computadores y Automática (2005)
2. Atienza, D., Mamagkakis, S., Catthoor, F., Mendías, J.M., Soudris, D.: Modular construction and power modelling of dynamic memory managers for embedded systems. In: Proceedings of International Workshop on Power And Timing Modeling, Optimization and Simulation (PATMOS). Lecture Notes in Computer Science (LNCS), vol. 3254, pp. 510–520. Springer, Berlin (2004). https://doi.org/10.1007/978-3-540-30205-6_53
3. Atienza, D., Mendías, J.M., Mamagkakis, S., Soudris, D., Catthoor, F.: Systematic dynamic memory management design methodology for reduced memory footprint. ACM Trans. Des. Autom. Electron. Syst. **11**(2), 465–489 (2006) https://doi.org/10.1145/1142155.1142165
4. Atienza Alonso, D., Mamagkakis, S., Poucet, C., Peón-Quirós, M., Bartzas, A., Catthoor, F., Soudris, D.: Dynamic Memory Management for Embedded Systems. Springer, Cham (2015). https://doi.org/10.1007/978-3-319-10572-7
5. Banakar, R., Steinke, S., Lee, B.S., Balakrishnan, M., Marwedel, P.: Scratchpad memory: a design alternative for cache on-chip memory in embedded systems. In: Proceedings of the International Symposium on Hardware/Software Codesign (CODES), pp. 73–78. ACM Press, Estes Park (2002). https://doi.org/10.1145/774789.774805
6. Bartzas, A., Peón-Quirós, M., Poucet, C., Baloukas, C., Mamagkakis, S., Catthoor, F., Soudris, D., Mendías, J.M.: Software metadata: systematic characterization of the memory behaviour

of dynamic applications. J. Syst. Softw. **83**(6), 1051–1075 (2010). https://doi.org/10.1016/j.jss. 2010.01.001. Software architecture and mobility

7. Berger, E.D., Zorn, B.G., McKinley, K.S.: Composing high-performance memory allocators. In: Proceedings of the ACM SIGPLAN Conference on Programming Language Design and Implementation (PLDI), pp. 114–124. ACM Press, Snowbird, Utah (2001). https://doi.org/10. 1145/378795.378821
8. Brassard, G., Bratley, T.: Fundamentals of Algorithmics, 1st (Spanish) edn, pp. 227–230. Prentice Hall, Englewood Cliffs (1996)
9. Chekuri, C., Khanna, S.: A PTAS for the multiple knapsack problem. In: Proceedings of the Annual ACM-SIAM Symposium on Discrete Algorithms (SODA), pp. 213–222 (2000). http:// dl.acm.org/citation.cfm?id=338219.338254
10. Clay Mathematics Institute (CMI): The millennium prize problems (2000). http://www. claymath.org/millennium-problems/millennium-prize-problems
11. Cohen, R., Katzir, L., Raz, D.: An efficient approximation for the Generalized Assignment Problem. Inf. Process. Lett. **100**(4), 162–166 (2006). https://doi.org/10.1016/j.ipl.2006.06.003
12. Cook, S.: The P versus NP problem (2000). http://www.claymath.org/sites/default/files/pvsnp. pdf
13. Cormen, T.H., Leiserson, C.E., Rivest, R.L., Stein, C.: Introduction to Algorithms, 2nd edn. MIT Press and McGraw-Hill Book Company (2001)
14. Gamma, E., Helm, R., Johnson, R., Vlissides, J.: Design Patterns: Elements of Reusable Object-Oriented Software. Addison-Wesley Longman Publishing, Boston (1995)
15. Grunwald, D., Zorn, B.: CustoMalloc: efficient synthesized memory allocators. Softw. Pract. Exp. **23**, 851–869 (1993). https://doi.org/10.1002/spe.4380230804
16. Jouppi, N.P.: Improving direct-mapped cache performance by the addition of a small fully-associative cache and prefetch buffers. In: Proceedings of the International Symposium on Computer Architecture (ISCA), pp. 364–373. ACM Press, Seattle (1990). https://doi.org/10. 1145/325164.325162
17. Kandemir, M., Ramanujam, J., Irwin, J., Vijaykrishnan, N., Kadayif, I., Parikh, A.: Dynamic management of scratch-pad memory space. In: Proceedings of the Design Automation Conference (DAC), pp. 690–695 (2001). https://doi.org/10.1145/378239.379049
18. Kandemir, M., Kadayif, I., Choudhary, A., Ramanujam, J., Kolcu, I.: Compiler-directed scratchpad memory optimization for embedded multiprocessors. IEEE Trans. Very Large Scale Integr. Syst. **12**, 281–287 (2004)
19. Lea, D.: A memory allocator (1996). http://g.oswego.edu/dl/html/malloc.html
20. McIlroy, R., Dickman, P., Sventek, J.: Efficient dynamic heap allocation of scratch-pad memory. In: Proceedings of the International Symposium on Memory Management (ISMM), pp. 31–40. ACM Press, Tucson (2008). https://doi.org/10.1145/1375634.1375640
21. Pisinger, D.: Algorithms for knapsack problems. Ph.D. thesis, University of Copenhagen (1995)
22. Poucet, C., Atienza, D., Catthoor, F.: Template-based semi-automatic profiling of multimedia applications. In: Proceedings of the International Conference on Multimedia and Expo (ICME), pp. 1061–1064. IEEE Computer Society Press, Silver Spring (2006)
23. Sindelar, M., Sitaraman, R., Shenoy, P.: Sharing-aware algorithms for virtual machine colocation. In: Proceedings of the ACM Symposium on Parallelism in Algorithms and Architectures (SPAA), pp. 367–378 (2011)
24. Soto, M., Rossi, A., Sevaux, M.: A mathematical model and a metaheuristic approach for a memory allocation problem. J. Heuristics **18**(1), 149–167 (2012). https://doi.org/10.1007/ s10732-011-9165-3
25. Steinke, S., Wehmeyer, L., Lee, B., Marwedel, P.: Assigning program and data objects to scratchpad for energy reduction. In: Proceedings of Design, Automation and Test in Europe (DATE), p. 409 (2002)
26. Verma, M., Steinke, S., Marwedel, P.: Data partitioning for maximal scratchpad usage. In: Proceedings of the Asia and South Pacific Design Automation Conference (ASP-DAC), pp. 77–83 (2003). https://doi.org/10.1145/1119772.1119788

27. Verma, M ., Wehmeyer, L., Marwedel, P.: Cache-aware scratchpad allocation algorithm. In: Proceedings of Design, Automation and Test in Europe (DATE) (2004)
28. Wilson, P.R., Johnstone, M.S., Neely, M., Boles, D.: Dynamic storage allocation: a survey and critical review. In: Proceedings of the International Workshop on Memory Management (IWMM), pp. 1–116. Springer, Berlin (1995). http://dl.acm.org/citation.cfm?id=645647.664690

Chapter 5
Design of a Simulator for Heterogeneous Memory Organizations

Our methodology for placement of dynamic data objects includes a step for simulation of memory hierarchies to evaluate the generated solutions and perform platform exploration. For example, if the design is at an early design phase and the actual platform is not yet available, the designer can use it to get an estimation of the placement performance. In systems such as FPGAs or ASICs in which the hardware can be modified, the designer can use the simulation to explore different architectural options. Alternatively, the simulator can also be used to estimate the performance of the applications on different models of a platform, either to choose the most appropriate one from a vendor or to steer the design of several models at different cost-performance points. After simulation, the designer can iterate on the previous steps of the methodology or, if the outcome looks satisfactory, use the output of the mapping step to prepare the deployment of the application.

In this chapter we will explore the design of a memory hierarchy simulator suitable for our methodology. Building our own simulator does not only enable easier integration with the rest of steps, but it offers also a context to explore the specific behavior of several memory technologies and how their characteristics affect the performance and energy consumption of the system. Understanding these peculiarities is essential to produce efficient data placements.

5.1 Overview

The simulator receives three inputs: the pool mappings, which include the correspondence between application IDs and memory resources, a trace of memory allocation operations and accesses, and the template of the memory hierarchy with annotated costs that was used during the mapping step. It then calculates the energy and number of cycles required for each access.

© Springer Nature Switzerland AG 2020
M. Peón Quirós et al., *Heterogeneous Memory Organizations in Embedded Systems*, https://doi.org/10.1007/978-3-030-37432-7_5

The simulator implemented in $\mathcal{D}yn\mathcal{A}s\mathcal{T}$ uses a model of the memory subsystem to simulate the behavior of the studied application and compute the energy and number of cycles required for each access. For example, DRAMs are modeled using the current, voltage, capacitive load, and timing parameters provided by the manufacturers. The simulator tracks the state of all the banks (which row is open in each of them) and includes the cost of driving the memory pins and the background energy consumed. These calculations are performed according to the rules stated by the JEDEC association—except for mobile SDRAM modules because they appeared before standardization and were developed independently by each manufacturer.

5.1.1 Elements in the Memory Hierarchy

The simulator that we are going to use in this chapter accepts descriptions of memory hierarchies in a hierarchical format. These templates can include any number of the following elements:

- **Multi-level bus structures.** The memory subsystem can be connected to the processor via a common bus, or through a multi-level hierarchy of buses with bridges between them. For each bus or bridge, it is possible to specify the energy and cycles that are required to pass through it.
- **Static RAMs.** SRAMs are the most basic type of memory for the simulator. Their main distinctive characteristic is that they are truly random-access memories: Accessing a word costs the same disregarding previous accesses to that module. As an example, scratchpad memories (SPMs) are usually implemented as SRAMs. The platform description can include SRAMs of any size, parameterized with their energy cost and latency per access.
- **Dynamic RAMs.** Main memory is usually implemented as DRAM due to its higher density and lower cost per bit in comparison with SRAM. In comparison with technologies such as Flash or other non-volatile memories, DRAMs are usually much faster, word-addressable and have a practically unlimited endurance. DRAMs are organized in banks, which has to be taken into account by the simulator to calculate the correspondence of addresses with rows and columns. In this chapter, we will study how to perform our calculations according to the rules for state transitions defined by the JEDEC association [2] and the manufacturer's datasheets. In particular, we will explore two types of DRAMs:
  - **Mobile SDRAM.** This low power mobile version, also known as LPSDR-SDRAM, transfers one data word on each clock cycle (e.g., on the positive edge), hence the denomination of single data rate. Multiple consecutive read or write accesses can be issued to random addresses in different banks; the memory outputs one word of data per cycle as long as the corresponding banks are active and no row misses happen. This technology allows burst sizes of 1, 2, 4 and 8 words or full-row.

- **LPDDR2-SDRAM.** Modern memories can transfer two data words per cycle, using both the rising and falling edges of the clock. The double data rate (DDR) denomination refers to this property. Consecutive read or write bursts can be executed to any number of banks without extra latency as long as all the banks are active and the appropriate rows are selected. A relevant feature is that most consecutive DDR standards increase the size of the internal DRAM pipeline to enable higher working frequencies. That means that short transfers may still incur some additional costs.

- **Cache memories.** Every DRAM module can have an associated cache hierarchy. Cache memories may be modeled according to several parameters: Size, associativity (from direct-mapped up to 16-ways), line length (4, 8 or 16 words), replacement policy (random or LRU, that is, least recently used), cached address range, latency and energy consumption per access. The simulator supports hierarchies of up to ten levels and each cache can cover the complete address range of a DRAM module, or just a part, enabling non-cached memory accesses. This last possibility enables the exploration of hybrid architectures in which some data elements with high locality are placed on cached DRAM, whereas others are managed explicitly on scratchpad memories or accessed directly from DRAM.

Although cache memories are not targeted by our data placement methodology, it is interesting to add support for them to the simulator for comparison purposes. Additionally, this also enables the exploration of hybrid memory hierarchies. The simulator produces results in CPU cycles. In the experiments we will assume that both SRAMs and caches work at the CPU frequency. Therefore, a memory latency of 1 cycle counts as one CPU cycle in the simulator. However, DRAMs work usually at a lower frequency than the CPU; hence, their latencies are multiplied by the factor between both frequencies during simulation.

5.1.2 DMM and Memory Address Translation During Simulation

The simulator reads the traces obtained during profiling and uses the memory allocation events, which contain the IDs of the corresponding dynamic data types (DDTs), to create and destroy representations of the objects mimicking the original execution of the application. This step is fundamental to reflect the behavior of the application because the memory accesses recorded in the traces reflected the placement of the data objects in the original memory hierarchy. As the memory organization used during profiling is different than the one used during mapping and simulation—after all, that is the very purpose of $DynAsT$!—the addresses assigned in the original execution of the application are meaningless in the context of any given simulation. Therefore, the simulator implements its own simple dynamic memory manager to reproduce on the memory space of the simulated platform each allocation recorded in the memory traces. Figure 5.1 illustrates the address translation mechanism.

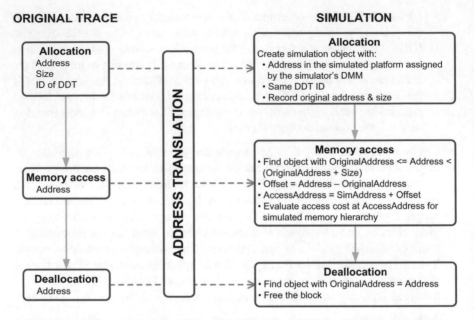

Fig. 5.1 Translation of the addresses recorded during profiling for use in the simulated platform. The simulator creates an object in the address space of the platform for every object allocated during the original execution and translates the address of each recorded memory access into the appropriate address for the simulated platform

Our simulator implements a memory allocator that always performs splitting and coalescing. Its execution speed does not affect the execution of the application as the DMM is run entirely by the simulator, out of the simulated platform's memory subsystem. In this way, memory accesses executed by the DMM used during simulation are not accounted. This can be changed by implementing in the simulator any concrete allocation algorithm, working directly on the memory space of the simulation. We can easily imagine new experiments such as the evaluation of the benefits of using a separate SRAM memory to hold the internal data structures used by the allocators. For example, it is not uncommon to avoid coalescing and splitting or full block searches in pools that are mapped in a DRAM to reduce the energy consumption of random accesses (which may force additional row activations). In contrast, if the data structures used by the DMM reside in a different small memory with a lower cost per access, more complex organizations for the DMM structures such as ordered trees instead of linked lists could be explored.

The memory allocator of the simulator uses two containers (implemented with std::multimap) to keep track of blocks: One, ordered by block size, for free blocks; the other, ordered by addresses, for used blocks. In that way, finding the best fit for a memory request is just a matter of traversing the tree of free blocks (with the method `std::multimap::lower_bound()`. Similarly, finding the block that corresponds to a deallocation operation consists in traversing the tree of used blocks, but using the request address instead of its size—which is usually not supplied to the deallocation function—as the search key.

Pools can be divided over several memory modules. In that case, the allocator tries first the heap in the most efficient memory. If it does not have enough space, the next heap is then probed. Finally, if none of the heaps of the pool has enough space, the memory request is forwarded to the backup pool. Objects are not reallocated, even if enough space becomes available in one of the other heaps of the pool, because of the same issues with data migration as outlined previously.

The simulator keeps also the active data instances (with an `std::map`), ordered by their address in the profiling trace. Each entry contains the ID of its DDT, its starting address in the original execution, its starting address in the simulated platform, and its size. When the simulator encounters a memory access in the profiling trace, it looks for the instance that covered that address in the original execution and calculates the offset of the access from the starting address of the object. Then, it uses the calculated offset and the starting address of the instance in the simulated platform to determine the exact address and memory module that corresponds to that access in the simulated memory hierarchy. In this way, the simulator correlates accesses recorded during profiling with the corresponding accesses during simulation. The IDs of the objects are used to count accesses to the different DDTs during simulation—the translated address identifies the memory resource and the pool of the accesses, but not the DDT of the object.

5.1.3   Life Cycle of a Memory Access in the Simulator

Figure 5.2 presents a high-level view of how memory accesses are evaluated by the simulator. For every memory access in the trace file, our simulator has to translate its address in the original platform to the corresponding address in the simulated platform. The new address identifies the affected memory module. In the case of SRAMs and uncached DRAMs, the corresponding memory model can be used directly to calculate the cost of the access. As we will see, this calculation is almost trivial in the case of SRAMs, whereas for DRAMs it involves tracking the state (e.g., active or idle) and open row of each bank. For cached ranges of DRAM, the simulator has to check if the accessed word is in the cache. If so, the cost of the access in the cache memory has to be calculated. On the contrary, if the word is not contained in the cache, the simulator has to evaluate the behavior of the complete memory hierarchy that intervenes in the movement, taking into account issues such as writing back modified lines that need to be evicted and data copies from as far as main memory to the first cache level.

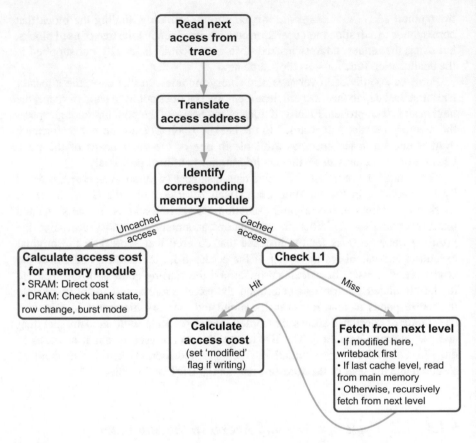

Fig. 5.2 Process followed by the simulator to calculate the cost of each memory access

We will make a few assumptions during the implementation of the simulator that are interesting to know. First, consecutive accesses to a memory word (or to bytes thereof) that may be registered by the profiling mechanisms are ignored because they are assumed to be handled inside the load/store queues of the processor. The processor is assumed not to access individual bytes through the system bus. Second, multi-cycle read accesses stall the processor because neither SRAMs nor caches are pipelined. However, DRAM writes do not necessarily stall the processor because they require one cycle in the bus per data word. Only if the memory is immediately used for a different operation the last write bears the complete latency. Finally, the simulation does not reflect delays between memory accesses due to long processing times (e.g., long chains of floating-point operations) as they are not recorded in the traces. However, most data-dominated applications are usually limited by memory bandwidth and latency rather than by processing time; therefore, this factor should not have a significant impact on the simulation outcome.

5.2 Simulation of Software-Controlled Scratchpads (SRAMs)

The behavior of static memories is the easiest to model. As they are truly random-access memories, the cost of reading or writing a word does not depend on the previous operations. In that sense, those memory modules do not have a notion of state.

Every SRAM in the platform template covers a range of addresses defined through a base address and a size. No other memory module's address range may overlap with the address range covered by a particular SRAM module. In that way, there is a unique correspondence between memory addresses and memory modules. Additionally, we will assume in the current implementation that SRAMs cannot be cached.

We will use four parameters to define an SRAM in a platform template: EnergyRead and EnergyWrite, which define the energy consumed during a read or a write access, respectively; and DelayRead and DelayWrite, which define the number of cycles required to complete a read or write access, respectively. The simulator simply accumulates the cost of each access to each SRAM. We will not consider more complex cases, such as pipelined write accesses for devices with latencies higher than 1 cycle.

5.3 Simulation of Cache Memories

Our methodology and our tool, $\mathcal{D}yn\mathcal{A}s\mathcal{T}$, are agnostic about caches: They are not considered when placing dynamic data objects into DRAMs, but the designer is free to include them either during simulation or in the actual hardware platform, as long as they do not interfere with other address ranges (specifically, with those covered by the SRAMs). However, the simulator supports the inclusion of cache memories in the platform templates to evaluate their effect on the remaining DRAM accesses or simply to compare our placement solutions with traditional cache-based systems (as we will do in Chap. 6).

We want to support in the simulator direct-mapped and associative caches, both with a configurable line size. Cache hierarchies can be "weakly" or strongly inclusive. In strongly inclusive cache hierarchies, every position contained in a level closer to the processor is guaranteed to be contained in the next level. For example, a word in L1 is also contained in L2. Exclusive cache hierarchies, which we will not support, guarantee that words are never contained in more than one level in the hierarchy. As an intermediate point, "weakly inclusive" (for lack of a standard term) hierarchies do not guarantee that a word in a level closer to the processor (e.g., L1) is also contained in the next level (e.g., L2). Enforcing complete inclusiveness eases maintaining cache coherence between multiple processors, but may introduce extra data movements if a line fetch in a further level produces an eviction that needs also to evict the corresponding line(s) in the closer levels, with the potential associated write-backs.

Fig. 5.3 Simulation of an access to a cached memory range. In the case of a cache miss, the simulator traverses the cache hierarchy as a real system would do, annotating energy consumption and latencies along the way

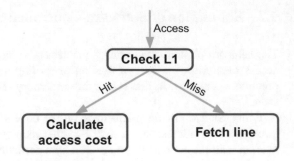

When the simulator processes a memory access from the trace file that corresponds to an object placed on a DRAM in the simulated platform, it has first to check if the address corresponds to a cached area to choose the corresponding DRAM or cache model. If a cached access results in a "cache hit," the associated energy and latency can be easily accounted. However, if it results in a "cache miss," then a recursive process has to be triggered to bring the cache line that contains that word from as far in the memory subsystem as necessary, accounting for potential evictions at each cache level. The simulator has to correctly account for the energy cost and latency of all the process (Fig. 5.3).

Figure 5.4 illustrates the simulation of line fetches. First, the simulator checks if the cache has to evict a modified line (unmodified lines can be discarded directly). If so, it starts a line write-back procedure. When the line can be safely reused, it checks if the cache is the last level, in which case it simply accounts for copying the data from the corresponding DRAM module. However, if the current cache level (l) is backed by other ones, the simulator has to recursively repeat the process for them: If the access is a hit for the next level ($l + 1$), the line is simply copied; but if it is a miss, then the address is recursively fetched starting at level $l + 1$.

Finally, Fig. 5.5 shows the process of writing a cache line to the next level in the hierarchy (write-back). Again, if the current cache level (l) is the last (or only) one, the line is written directly to main memory. The simulator invalidates automatically lines that are written back, so that they can be reused. If the current cache level is backed by another one, the simulator checks whether the cache at $l + 1$ contains a copy. If the line is present at level $l + 1$, the updated contents are simply passed from level l to level $l + 1$. Otherwise, the line is copied to $l + 1$, with the exception that if it has longer lines, then the complete line must be first fetched there to ensure a correct partial update. Finally, if level $l + 1$ does not have a free line to accommodate the line from level l—which may happen in the case of weak inclusion—then a write-back procedure is recursively started to free one cache entry at level $l + 1$.

Figures 5.4 and 5.5 can be modified to force complete inclusion as follows. When a cache line is going to be overwritten at cache level l (either to receive data from main memory or because of the invalidation after a write-back), the simulator checks if the same address is also present at level $l - 1$. If so, the line is also purged (written back if needed) from $l - 1$. This extra work guarantees that an address is not present at a level if it is not also present at the next ones.

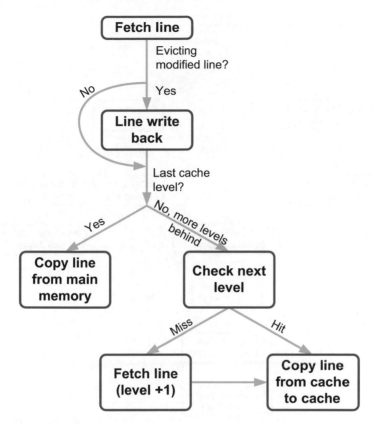

Fig. 5.4 Simulation of cache line fetch. In multi-level cache hierarchies, the process is potentially recursive until the main memory, usually a DRAM, is reached

Checking whether an address is present at a cache level counts as an access: misses also incur some energy and latency costs. This is because caches do commonly read the tag and data arrays in parallel, which improves hit access time. A possible extension, particularly interesting in the domain of low power embedded systems, would be to consider that the cache reads just the tag array and then, in case of a hit, the actual data contents, which can help to reduce cache energy consumption, particularly for caches with high associativity.

One interesting remark is that we do not need to store the actual data in the simulator: The simulator only accounts for energy consumption and latency, and sets the state of each cache line (address tag, and validity and modification bits).

5.3.1   Overlapped Accesses

Transfer operations between cache levels or main memory are normally pipelined, which reduces the total latency. Therefore, the simulator assumes that words

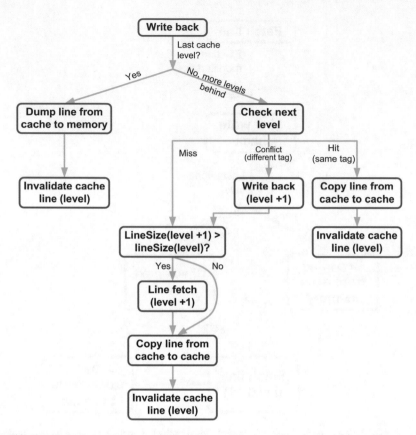

Fig. 5.5 Simulation of cache line write-back. As in the case of line fetches, in multi-level cache hierarchies the process is potentially recursive until the main memory is reached

are transferred one at a time at the pace of the slowest memory. However, we will not add support for "hot-word first" operation—as processors themselves are not simulated, it would be difficult to assess the benefits of this technique in the execution pipeline—nor for multiport caches. We will calculate the energy consumption of transferring complete lines as the addition of the energy required to transfer individual words.

5.3.2   Direct-Mapped Caches

Direct-mapped caches are the simplest type of cache memories. The cacheable address space is divided in blocks of the same size than the cache line (e.g., 64 B for lines containing 16 words of 32 bits). A mapping function is used to assign each line in the address space to a line in the cache memory. The simplest case, and also

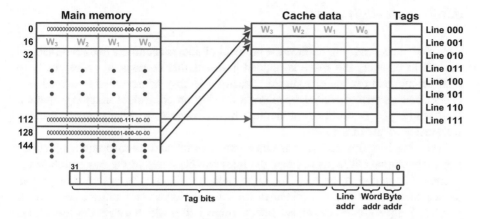

Fig. 5.6 Direct-mapped cache with 8 lines of 4 words each (total cache size is 128 B). Memory addresses alias (collide) every 128 B on the same cache line. The cache controller checks the tag bits stored with each line against the access' address to discriminate between hits and misses

the most commonly used, divides the address bits into fields for byte, word-in-line and line, and then uses the line bits to address directly into the cache memory.

This simple mapping of the complete address range into the address range of the cache memory produces conflicts for addresses separated exactly by a multiple of the size of the cache memory. In any case, no matter what mapping scheme is used, conflicts are inevitable (by the pigeonhole principle). To identify them, the cache memory stores the upper part of the address ("address tag") along the data values. Thus, the cache knows implicitly part of the address of each line contents and stores explicitly the rest of the address bits. The combination of line addressing and tag comparison allows the cache controller to determine if an address is contained in a cache. Figure 5.6 shows this simple scheme.

To implement direct-mapped caches in our simulator, we need as parameters the total cache capacity and the number of words per line. Then, we can calculate automatically the number of bits used for word addressing, the number of lines in the cache, and the number of tag bits with Eqs. (5.1)–(5.5).

$$byteAddrSize = \log_2 \overbrace{bytesPerWord}^{4} \tag{5.1}$$

$$wordAddrSize = \log_2 wordsPerLine \tag{5.2}$$

$$numLines = (cacheSize/bytesPerWord)/wordsPerLine \tag{5.3}$$

$$lineAddrSize = \log_2 numLines \tag{5.4}$$

$$tagSize = \underbrace{addrSize}_{32} - wordAddrSize - lineAddrSize - \underbrace{byteAddrSize}_{2} \tag{5.5}$$

5.3.3   Associative Caches

Associative caches try to palliate the problem of address conflicts in direct- mapped caches. In essence, the space is divided in n sets that correspond conceptually to n direct-mapped caches of reduced capacity. Memory addresses are mapped into cache lines as with direct-mapped caches, using an equivalent mapping function (frequently, just a subset of the address bits). However, a given memory address may reside in any of the n sets.

The benefit of this approach is that circumstantial address-mapping conflicts do not force mutual evictions because the lines can be stored at the same time in the cache, each on one of the sets. The disadvantage is that all the sets must be probed during each access and a mechanism for set selection when storing a new line is needed. Frequent choices for set selection are LRU, which evicts the line in the least recently used set with the hope that it will not be needed soon, and random replacement.

Designs with two and four sets (also known as "ways") tend to produce the biggest improvements in comparison with direct-mapped caches, trading between hit rate and cost per access (because of the multiple tag comparisons), while higher degrees of associativity tend to provide diminishing returns. Figure 5.7 illustrates the structure of an associative cache with two sets.

The different sets in an associative cache are not identified by addressing bits. Instead, the full tag is stored for each set and compared when searching for a line. Additionally, the number of lines is divided by the number of sets, in comparison with a direct-mapped cache.

To implement associative caches in our simulator, we need as parameters the total cache capacity, the number of words per line, and the number of sets (ways).

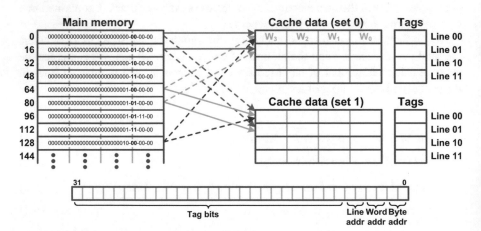

Fig. 5.7 2-Way associative cache with 4 lines of 4 words each (total cache size is 128 B). Every 64 B, memory addresses collide on the same cache lines, but they can be stored on any of the two sets. The cache controller checks the tag bits stored in each set for the corresponding line against the access' address to discriminate between hits to any of the sets and misses

Then, we can calculate automatically the number of bits used for word addressing, the number of lines in the cache, and the number of tag bits with Eqs. (5.6)–(5.10).

$$byteAddrSize = \log_2 \overbrace{bytesPerWord}^{4} \tag{5.6}$$

$$wordAddrSize = \log_2 wordsPerLine \tag{5.7}$$

$$numLines = ((cacheSize/bytesPerWord)/wordsPerLine)/numSets \tag{5.8}$$

$$lineAddrSize = \log_2 numLines \tag{5.9}$$

$$tagSize = \underbrace{addrSize}_{32} - wordAddrSize - lineAddrSize - \underbrace{byteAddrSize}_{2}$$
$$\tag{5.10}$$

5.4 Overview of Dynamic Memories (DRAMs)

Accurate simulation of DRAM modules is a complex and error-prone process. Therefore, here we explore as many details of the simulator implementation as reasonable so that readers of this text can evaluate themselves the conditions on which the experiments of Chap. 6 were performed.

5.4.1 Why a DRAM Simulator

Manufacturers such as Micron provide simple spreadsheets to estimate the average power required by a DRAM during the whole execution of an application. However, calculating energy consumption requires knowing also the execution time ($E = P \times t$). The main factors that affect DRAM performance are the number of row changes,[1] the exact combination and number of switches between reads and writes, and the number of accesses that are executed in burst or individually (the latency of the first access is amortized over the length of a burst). In our simulator, we have to track the current state of each memory bank to calculate accurately the number of cycles that each memory access lasts and compare the performance of different solutions.

An additional interesting advantage of implementing a DRAM simulator that reflects energy consumption on an operation basis is that the designer can iden-

---

[1] The worst case happens when every operation accesses a different row in a single bank. Assuming that t_{RC} is 10 cycles and t_{RCD}, t_{RP}, and CL are 3 cycles in an SDRAM operating at 166 MHz, each memory access lasts 10 cycles ($t_{RC} \geq t_{RCD} + t_{RP} + CL$). Thus, only a 10% of the maximum bandwidth available in burst mode is reached.

tify peaks of energy consumption that affect negatively system performance or stability. For example, our methodology could be extended to include temperature calculations based on cycle-accurate energy consumption and track both spatial and temporal temperature variations.

Last but not least, the design of a DRAM simulator such as the one that we are studying here provides a context to understand the particularities of the behavior of different DRAM technologies and their impact on performance and energy consumption.

5.4.2   DRAM Basics

Main memory is usually implemented as DRAM due to its higher density and lower cost per bit. DRAMs present a good compromise between the performance of SRAMs and the density of non-volatile memory technologies such as Flash or resistive memories. Both SRAMs and DRAMs have an endurance that can be considered, for all practical purposes, infinite. However, in comparison with SRAMs, the elements in the DRAM cell array cannot be accessed directly. Memory cells are grouped in rows of usually 1024–8192 words (Fig. 5.8). At the memory module level, an internal (SRAM) buffer latches the contents of one active row. Words in the active row buffer can be accessed efficiently.

To access a different row, the module undergoes a two-step process: First, the pairs of bit lines that traverse the columns in the cell array are "precharged" at

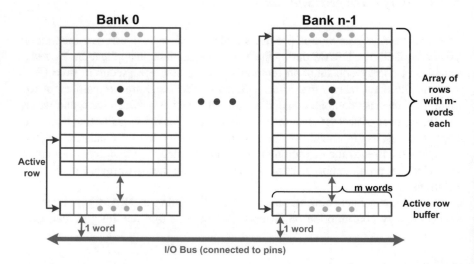

Fig. 5.8 The cells in the DRAM matrix cannot be accessed directly. Instead, DRAMs are divided in several banks and each one has a buffer that holds at most an active row. Only the words in the active row buffer can be accessed directly

specific voltage values. Then, the transistors of the cells in the selected row are "activated" to connect the cells to one of the bit lines. The sense amplifiers detect and amplify the voltage difference between the two bit lines. The read values are latched in the row buffer so that the individual words can be accessed efficiently. The operation of the sense amplifiers "refreshes" the cells that have been read, but this operation requires some time that must be respected between an activation and the next precharge.

Additionally, as DRAMs are based on capacitors that lose their charge over time, the cells in every row have to be "refreshed" periodically. As a result, the DRAM becomes inaccessible every few microseconds as the controller refreshes one row each time (which basically means precharging the bit lines and activating that row). The time that the rows can endure before losing their data diminishes as temperature increases; hence, this is another good reason to save energy and keep the temperature of the device low.

To reduce the need for switching active rows, DRAM modules are divided into several (usually 4–16) banks. Each bank has one row buffer, that is, at most one row can be active at a time in each bank. Words from the open rows in any of the banks can be accessed seamlessly, but changing the active row in a bank incurs an extra cost—however, oftentimes a careful block and row management can hide some of the latencies by reading data from a bank while other is preparing a new row. Therefore, the cost of an access depends on the currently active row for the corresponding bank, which is a crucial difference with the working of SRAMs. Interestingly, Marchal et al. [3] already showed how a careful allocation of data to DRAM banks can reduce the number of row misses and improve latency and energy consumption.

Figure 5.9 presents a generic (and simplified) diagram of DRAM bank states. The default state for the banks of a DRAM is *IDLE*, where no row is open. When an access needs to be performed, the memory controller has first to activate the row that corresponds to the memory address. This process moves the bank to the *ACTIVE* state, copying the data corresponding to the required row from the main DRAM array to the active row buffer. While a row is in the buffer, any number

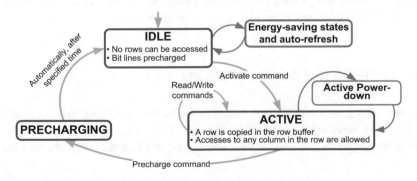

Fig. 5.9 Simplified state diagram for the banks of a DRAM

of read or write accesses, in any order, can be performed to any of its columns. When the controller needs to access a word in a different row, it must issue a PRECHARGE command to the bank. The bank remains in the *PRECHARGING* state for a specified period of time and, after that, the controller can issue the activation command needed to access the new row and move the bank to the *ACTIVE* state again. The JEDEC association publishes a standard regulating the logical behavior of each DRAM technology and the minimum physical parameters (timing, voltages, currents, etc.) that the manufacturers must respect to claim compliance with it.

In general, accesses to the active rows of any of the banks can be alternated efficiently in almost every type of DRAM. In particular, the DRAMs that we are going to implement in our simulator support continuous bursts of reads or writes. Burst modes are active for a number of words; individual or burst accesses can be linked if the respective commands are presented on the command lines of the bus at the appropriate times. Accessing a word in a row that is not active requires precharging the bit lines and activating that row.

A flag on every access command allows the memory controller to specify if the row should be closed (i.e., a PRECHARGE automatically executed) after it. Although in our simulator we are going to implement an "open-row" policy, that is, rows are kept active until an access forces a change, other policies can be explored. For example, in the "closed-row" policy the controller closes the active row after every burst; however, this technique can incur significant increases on energy consumption and instantaneous power requirements (currents in the circuit are much higher during activations), which should be avoided in the design of low power embedded systems. Other intermediate possibilities close the active rows if they are not used after a certain time, which allows the bank to enter a lower-power mode. An interesting option would be modifying the simulator to assume that the memory controller has enough information to guess if a row should be closed during the last access to the previous row. Thus, given enough time between accesses to different rows, the delays could be practically hidden (of course, energy would be equally consumed). A similar technique seems to have been applied by some high-end processors [1].

DRAM timings (e.g., CL, t_{RCD}, or t_{Read}) are expressed in terms of bus cycles. As our simulator works with CPU cycles, all timing parameters are multiplied by *CPUToDRAMFreqFactor* during simulator initialization.

5.5   Simulation of Mobile SDRAMs

For this model, we will follow Micron's datasheets for their Mobile LPSDR-SDRAM [4], whose most fundamental characteristic is being a pipelined architecture instead of a prefetch one. In the datasheet we can find the following remarks in this regard:

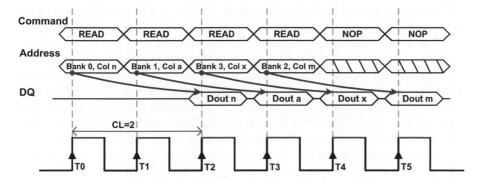

Fig. 5.10 "Seamless burst" access in an SDRAM with $CL = 2$, random read accesses. Each READ command can be issued to any bank as long as they are active. Successive READ commands can be issued on every cycle. The result of each operation is available CL cycles after the command is issued. This example shows linked bursts of length 1; longer burst sizes retrieve several consecutive data words with a single READ command. In that case, continuous reading can be achieved issuing each READ command CL cycles before the preceding burst finishes

> Mobile LPSDR devices use a pipelined architecture and therefore do not require the $2n$ rule associated with a prefetch architecture. A READ command can be initiated on any clock cycle following a READ command. Full-speed random read accesses can be performed to the same bank, or each subsequent READ can be performed to a different bank. [4, p. 43]

And

> Each READ command can be issued to any bank. [4, p. 44 Note 1; p. 45 Note 1]

Similar remarks are made for write accesses in subsequent pages of the datasheet.

With the previous information, we can assume in the simulator that all consecutive (in time) read accesses form a burst; in particular, bursts can include reads to any words in the set of active rows, no matter their absolute addresses (Fig. 5.10). The first access in a burst reflects the complete delay of the pipeline (CL), but the memory outputs one word in each subsequent bus cycle (t_{Read} is normally one cycle). Energy consumption is calculated for the whole duration of the access (CL for the first, 1 for the next) because I_{DD4} is measured during bursts and thus, it reflects the energy consumed by all the stages of the pipeline. A possible error source is that I_{DD4} might be slightly lower during the CL cycles of the first access as the operation is progressing through the pipeline, but there seems to be no available data in this respect.

A burst is broken by an inactivity period or an access to a different row. In those cases, the first access of the next burst has to reflect the complete latency. Inactivity periods (i.e., after the last word of a burst is outputted) happen if the processor accesses other memories for a while; for instance, after several consecutive accesses to an internal SRAM or if cache memories are effective and thus DRAMs are seldom accessed. Accesses to words of a non-active row on any bank break a burst and thus the simulator has to account for corresponding the precharge and activation times. The simulator could be extended assuming that the PRECHARGE command was sent to the affected bank as far back in time as the last access to that bank (every access command may include an optional PRECHARGE operation) and, therefore, simulate interleaving of bank accesses with no intermediate latencies.

Writes are only slightly different. Once starting delays are met (e.g., after a row activation), the pipeline does not introduce additional delays for the first access; each data word is presented on the bus during one cycle (t_{Write} is 1 in the used datasheets). However, the banks require a "write-recovery time" (t_{WR}) after the last write before a PRECHARGE command can be accepted. Every time that the simulator has to change the active row in a bank, it has to check if that bank is ready to accept new commands by evaluating $t_{LastBankWrite} + t_{WR} \geq CurrentTime$. Energy consumption is calculated for one cycle (t_{Write}) for normal write accesses; however, when a burst is finished (due to inactivity) or the activity switches to a different bank, the extra energy consumed by the bank finishing its work in the background during t_{WR} cycles needs also to be accounted.

Our simulator does not implement any policy for proactively closing rows and thus does not currently account for the potential energy savings. This topic is worth future expansion.

Address organization in a DRAM module is a matter of memory controller design. A common choice is to organize address lines as "row-bank-column" so that logical addresses jump from row j in bank m to row j in bank $m + 1$. This option is beneficial for long sequential accesses because the memory controller can precharge and activate the row in the next bank on advance, so that when the words in the active row of the current bank are read, it can continue immediately with the next bank. However, we will assume that address lines are organized as "bank-row-column": Addresses cover first a complete bank, then continue in the next one. The downside of this choice is that long sequential accesses that spawn several rows must wait for the precharge and activation times after exhausting each row. However, it enables pool placement on DRAM banks because the banks can be seen as continuous ranges of memory addresses.

Example 5.1 Addressing in a 256 MB 4-bank DRAM.

With our organization of the memory space (bank-row-column), each DRAM bank is seen as a continuous range of 64 MB. Assuming each row has 1024 words, the application can access 4 KB of data before encountering a precharge-activate delay, which represents a relatively small penalty. This penalty is somewhat higher in the case of sequential access patterns for very small data objects that cross the boundary between rows.

In comparison, with a row-bank-column organization, DMM pools would have to be organized in 1024-word heaps, being impossible to allocate objects across different heaps (i.e., rows) without crossing to other banks. Therefore, each heap would have to be completely independent and serious fragmentation issues might appear, affecting not only to objects bigger than 4 KB. Additionally, the system would suffer the complexity of managing thousands of heaps, one per row in each bank.

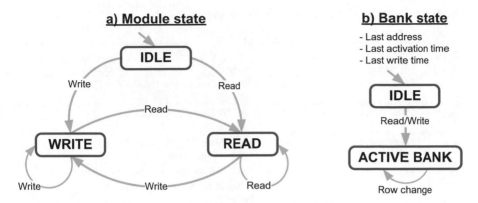

Fig. 5.11 Module and bank state in the simulator. All transitions check if a change of active row is needed. Transitions from write that require a row change have to observe t_{WR}

In any case, the effect of different address space organizations is open to further investigation. For example, Zhang et al. [6] presented an address organization based on permutations that reduces the number of row conflicts due to L2-cache misses.

Finally, in the simulator we will ignore the effect of DRAM refreshing on energy consumption and row activations. The datasheet of the modeled device mandates a refresh command to each row in a bank every 64 ms, which can be spread as a one-row refresh every 7.8125 μs or 8192 row refreshes clustered at the end of the 64 ms period (t_{REF}). The duration of a row refresh is specified by the t_{RFC} parameter (72 ns for the modeled device). The total accumulated time spent refreshing rows is $8192 \times 72\,\text{ns}/64\,\text{ms} \approx 0.9\%$, which represents a small percentage of the total. Similarly, it usually represents less than 0.1% of the total energy consumption.

Given the previous assumptions, the simulator has to track (Fig. 5.11):

- For the complete module, the state (reading or writing) and the last access time (to know if new accesses extend the current burst or start a new one).
- For each bank, its active row, the time of the last write (to account for the write recovery time) and the time of the last row activation (to respect t_{RAS} and t_{RC}, Table 5.1).

The simulator does not need to track the specific read or write state of each bank, only if a different row needs to be activated. This is because I_{DD4} is measured for bursts of reads or writes to any bank and has the same value for both types of accesses in the datasheets.

5.5.1 Memory Working Parameters

Table 5.1 defines the parameters needed in the simulator to model SDRAMs. Timing parameters are usually provided by the manufacturers in nanoseconds (ns); system designers round them up to DRAM bus cycles.

Table 5.1 Definition of standard working parameters for SDRAMs

Name	Units	Typical	Description
t_{CK}	ns	6	Clock cycle time
CL	Bus cycles	3	CAS (column-address strobe) latency
t_{CDL}	Bus cycles	1	Last data-in to new READ/WRITE command
t_{RAS}	Bus cycles	7	ACTIVE-to-PRECHARGE command
t_{RC}	Bus cycles	10	ACTIVE-to-ACTIVE command period (same bank)
t_{RCD}	Bus cycles	3	ACTIVE-to-READ-or-WRITE delay (**row** address to **c**olumn address **d**elay)
t_{RP}	Bus cycles	3	PRECHARGE command period
t_{WR}	Bus cycles	3	Write recovery time
I_{DD0}	mA	67.6	ACTIVE-PRECHARGE current (average, calculated)
I_{DD1}	mA	90.0	Operating current
I_{DD3}	mA	18.0	Standby current, all banks active, no accesses in progress
I_{DD4}	mA	130.0	Operating current, read/write burst, all banks active, half data pins change every cycle
V_{DD}	V	1.8	Supply voltage
V_{DDq}	V	1.8	I/O supply voltage (usually, $V_{DDq} = V_{DD}$)
C_{L0}	pF	20.0	Input/output pins (DQs) capacitance (for input, i.e., writes)
C_{LOAD}	pF	20.0	Capacitive load of the DQs (for output, i.e., reads)

Timing parameters are usually provided in ns and rounded up to DRAM bus cycles

CL (CAS or "column-address strobe" latency) is the latency since a READ command is presented and the data are outputted. t_{RP} and t_{RCD} determine the time since a bank is precharged until the new row is active and ready to be accessed. t_{WR} is counted when the active row of a bank is changed to ensure that the last write succeeds. t_{RC} defines the minimum time between two row activations in the same bank. t_{RAS} defines both the minimum time between an ACTIVATE command and the next PRECHARGE to the same bank, and the maximum time that a row may remain active before being precharged. The simulator observes implicitly the minimum period for t_{RAS} as normally $t_{RC} = t_{RP} + t_{RAS}$. The maximum period, which forces to reopen a row periodically, is ignored as typical values are in the order of 120,000 ns (we assume in the simulator that rows may stay active indefinitely).

I_{DD4}, the current used during bursts, is calculated assuming that "address transitions average one transition every two clocks." This value applies for random reads or writes from/to any bank with no other banks precharging or activating rows at the same time.

5.5.2 Calculations

The following paragraphs explain how to calculate some derived quantities in our
simulator.

5.5.2.1 IDD_0: Maximum Operating Current

I_{DD0} is normally defined as the average current during a series of ACTIVATE
to PRECHARGE commands to one bank. Micron's power estimation spreadsheet
calculates it as follows:

$$I_{DD0} = I_{DD1} - 2 \frac{t_{CK}}{t_{RC}} (I_{DD4} - I_{DD3}) \qquad (5.11)$$

5.5.2.2 Shortcuts

We will use the following shortcuts to simplify the writing of long equations:

$$P_{ActPre} = I_{DD0} \times V_{DD} \qquad (5.12)$$

$$P_{Read} = I_{DD4} \times V_{DD} \qquad (5.13)$$

$$P_{Write} = I_{DD4} \times V_{DD} \qquad (5.14)$$

$$t_{CPUCycle} = 1/CPUFreq \qquad (5.15)$$

P_{ActPre} (5.12) gives the power used during precharge and activation operations.
As I_{DD0} is measured as an average over consecutive pairs of both commands, P_{ActPre}
is also an average. In reality, the power required during precharging is much less
than during a row activation; however, as both commands come in pairs, the overall
result should be an adequate approximation.

P_{Read} (5.13) and P_{Write} (5.14) represent the power required during a burst of
reads or writes, respectively. $t_{CPUCycle}$ is calculated (5.15) using the CPU frequency
defined in the platform template file and is normally measured in ns.

We will also use t_{Read} as a shortcut to represent the number of bus cycles that read
data are presented by the memory module on the external pins. Correspondingly,
t_{Write} represents the number of bus cycles that written data must be presented by the
memory controller on the external data pins. Normally, both values are one cycle:
$t_{Read} = t_{Write} = 1$.

5.5.2.3 Power to Drive Module Pins

Equations (5.16) and (5.17) give the power required to drive the module pins
during read or write accesses, respectively. The energy required for a complete

$0 \rightarrow 1 \rightarrow 0$ transition corresponds to $C \times V_{DDq}^2$. Since the signals toggle at most once per clock cycle, their effective frequency is at most $0.5 \times CPUFreq/CPUToDRAMFreqFactor$.[2]

$$PDQ_r = \underbrace{32}_{\text{32 pins}} \times 0.5 \times \underbrace{C_{LOAD} \times V_{DDq}^2}_{\text{Transition energy}} \frac{CPUFreq}{CPUToDRAMFreqFactor} \qquad (5.16)$$

$$PDQ_w = \underbrace{32}_{\text{32 pins}} \times 0.5 \times \underbrace{C_{L0} \times V_{DDq}^2}_{\text{Transition } E} \frac{CPUFreq}{CPUToDRAMFreqFactor} \qquad (5.17)$$

To calculate the power needed for writes, we will assume that the capacitive load supported by the memory controller (C_{L0}) is the same than the load driven by the memory during reads: $C_{L0} = C_{LOAD}$. Although this does not strictly correspond to energy consumed by the memory itself, it is included as part of the energy required to use it.

Finally, energy consumption is calculated multiplying P_{DQ} by the length of an access. Alternatively, the simulator could simply use $E = 32 \times C \times V_{DDq}^2$ and multiply by the number of (complete) transitions at the data pins.

5.5.2.4 Background Power

The Micron datasheets used as reference calculate the power required by each operation subtracting the current used by the module in standby mode, I_{DD3} (with all the banks active), from the total current used: $P_{ActPre} = (I_{DD0} - I_{DD3})V_{DD}$ and $P_{Read} = P_{Write} = (I_{DD4} - I_{DD3})V_{DD}$. Then, the background power (the power required just by having all the modules active without doing anything) is calculated apart.

However, in our simulator we will calculate directly the total energy consumed during these operations, using I_{DD0} and I_{DD4}. The energy consumed during real standby cycles will be calculated later (5.18) using the total number of standby cycles counted by the simulator (i.e., cycles during which the DRAM banks were active but the DRAM was not responding to any access). We believe that both approaches are equivalent.

$$E_{Background} = I_{DD3} \times V_{DD} \times EmptyCycles \times t_{CPUCycle} \qquad (5.18)$$

---

[2]The amount of transitions at the DRAM pins is data-specific. A better approximation could be achieved by including during profiling the actual data values read or written to the memories, at the expense of a bigger log file size.

5.5.3 Simulation

The simulation can be organized according to the state of the module and the type of the next operation executed. Every time that a row is activated, the simulator has to save the operation time to know when the next activation can be initiated. Similarly, the simulator tracks the time of the last write to each bank to guarantee that the write-recovery time (t_{WR}) is met.

The following diagrams present all the timing components required during each access. Some parameters limit how soon a command can be issued, and are counted since a prior moment. Thus, they may have already elapsed at the time of the current access. These parameters are surrounded by square brackets ("[]"). This applies particularly to t_{RC}, which imposes the minimum time between two ACTIVATE commands to a bank required to guarantee that the row is refreshed in the cell array after opening it.

The simulator identifies periods of inactivity for the whole module, providing the count of those longer than 1000 CPU cycles and the length of the longest one. This information may be used to explore new energy-saving techniques.

5.5.3.1 From the IDLE State

The transition from the *IDLE* state happens in the simulator's model only once, at the beginning of every DRAM module simulation (Fig. 5.12).

READ from IDLE
The banks are assumed to be precharged, so that the access has to wait for the row-activation time (t_{RCD}) and the time to access a word in the row buffer (CL). Equation (5.19) gives the energy required to complete the access. E_{Read} is calculated for the whole CL time to reflect the latency of the first access in a burst. As data are presented on the bus for just one cycle, the energy consumed driving external pins is confined to that time, represented by t_{Read}.

$$
\begin{aligned}
E &= E_{Activation} + E_{Read} + E_{DrivePins} \\
&= (\underbrace{P_{ActPre} \times t_{RCD}}_{\text{Activation}} + \underbrace{P_{Read} \times CL}_{\text{Read}} + \underbrace{PDQ_r \times t_{Read}}_{\text{Drive pins}}) \times t_{CPUCycle}
\end{aligned}
\tag{5.19}
$$

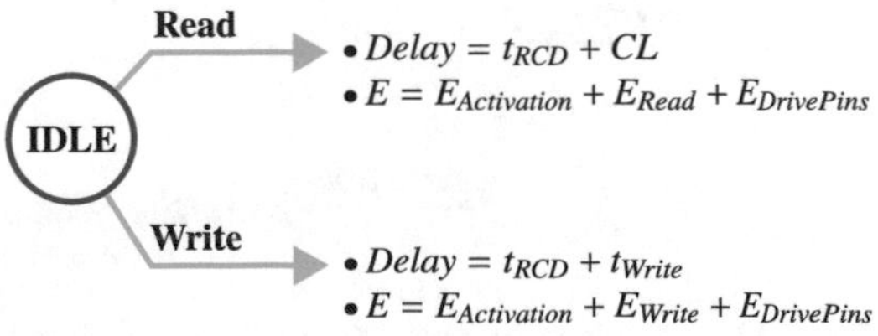

Fig. 5.12 Initial transition for an LPSDRAM module

Read
- $Delay = t_{RCD} + CL$
- $E = E_{Activation} + E_{Read} + E_{DrivePins}$

IDLE

Write
- $Delay = t_{RCD} + t_{Write}$
- $E = E_{Activation} + E_{Write} + E_{DrivePins}$

t_{RCD}, CL, and t_{Read} are measured in bus cycles, whereas $t_{CPUCycle}$ is the length in seconds of each cycle. Ergo, each of t_{RCD}, CL, and t_{Read} multiplied by $t_{CPUCycle}$ expresses a time period measured in seconds.

WRITE from IDLE

Similarly, a WRITE access has to wait until the required row is active and then one extra cycle (t_{Write}) during which data are presented on the bus to the memory module. Equation (5.20) gives the energy required to complete the write. The memory controller consumes energy driving the external pins during one cycle (t_{Write}).

$$E = E_{Activation} + E_{Write} + E_{DrivePins}$$
$$= (\underbrace{P_{ActPre} \times t_{RCD}}_{\text{Activation}} + \underbrace{P_{Write} \times t_{Write}}_{\text{Write}} + \underbrace{PDQ_w \times t_{Write}}_{\text{Drive pins}}) \times t_{CPUCycle} \qquad (5.20)$$

5.5.3.2   From the READ State

Figure 5.13 shows the possible transitions from the READ state.

READ After READ with Row Change

A READ command that accesses a different row than the one currently active requires a full PRECHARGE-ACTIVATE-READ sequence. Additionally, a minimum of t_{RC} cycles

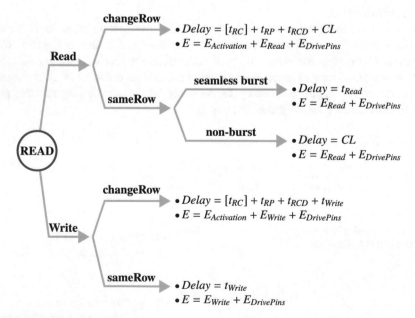

Fig. 5.13 Transitions after previous read accesses (LPSDRAM)

must have elapsed since the last ACTIVATE command to that bank. Therefore, the simulator has to check the last activation time for the bank and calculate the remaining part of t_{RC} that still needs to pass (under most normal conditions, it will be zero). Equation (5.21) shows how the components of the energy consumption of this operation are calculated:

$$E = (\overbrace{P_{ActPre} \times (t_{RP} + t_{RCD})}^{\text{Activation}} + \overbrace{P_{Read} \times CL}^{\text{Read}} + \overbrace{PDQ_r \times t_{Read}}^{\text{Drive pins}}) \times t_{CPUCycle}$$

$$(5.21)$$

READ After READ in the Active Row

A READ command that accesses a word in the active row can proceed directly without more delays. However, the simulator makes a distinction between access that are consecutive in time and accesses that are separated by a longer time.

In the first case, the simulator assumes that the READ command belongs to the previous burst access. As the time to fill the pipeline (CL) was already accounted during that access, the delay for the current access is one cycle (t_{Read}). In other words, the memory controller will receive the data after CL cycles, but the delay of this access with respect to the previous is only one cycle. Figure 5.10 presented this case: The first read command at T_0 was answered at T_2; however, the accesses started at T_1, T_2, and T_3 received their data consecutively at T_3, T_4, and T_5. Equation (5.22) shows the detailed equation for energy consumption:

$$E = (\overbrace{P_{Read} \times t_{Read}}^{\text{Read}} + \overbrace{PDQ_r \times t_{Read}}^{\text{Drive pins}}) \times t_{CPUCycle}$$

$$(5.22)$$

In the second case, the current access is the first of a new burst. Therefore, the simulator accounts the full pipeline delay to it, CL. Equation (5.23) details energy consumption in this case:

$$E = (\overbrace{P_{Read} \times CL}^{\text{Read}} + \overbrace{PDQ_r \times t_{Read}}^{\text{Drive pins}}) \times t_{CPUCycle}$$

$$(5.23)$$

WRITE After READ with Row Change

A WRITE command that accesses a row other than the currently active one starts a full PRECHARGE-ACTIVATE-WRITE cycle. Similarly to other cases, the simulator has to check whether the minimum t_{RC} time between to ACTIVATE commands has already been met or not. Equation (5.24) shows the details of energy consumption:

$$E = (\overbrace{P_{ActPre} \times (t_{RP} + t_{RCD})}^{\text{Activation}} + \overbrace{P_{Write} \times t_{Write}}^{\text{Write}} + \overbrace{PDQ_w \times t_{Write}}^{\text{Drive pins}}) \times t_{CPUCycle}$$

$$(5.24)$$

WRITE After READ in the Active Row

WRITE commands do not wait for a result, so they take one bus cycle as long as they access one of the active rows. Equation (5.25) shows the details of energy calculation in this case:

$$E = (\overbrace{P_{Write} \times t_{Write}}^{Write} + \overbrace{PDQ_w \times t_{Write}}^{Drive\ pins}) \times t_{CPUCycle} \quad (5.25)$$

5.5.3.3 From the WRITE State

Figure 5.14 shows the possible transitions from the WRITE state.

READ After WRITE with Row Change

A READ command after a WRITE that changes the active row has two peculiarities. First, before the PRECHARGE command can start, the minimum write-recovery time (t_{WR}) must be respected. Second, the new row cannot be activated until a minimum t_{RC} time since the previous ACTIVATE command has elapsed. Therefore, the simulator checks both times and delays the new READ operation until both times have been met. From this point, the read proceeds normally as in previous cases.

Equation (5.26) details how energy consumption is calculated in this case. Write accesses are accounted by the simulator in the reverse way than read accesses: Each of them has a delay of one cycle (t_{Write}), except the last one, which requires a few extra cycles to complete. As I_{DD4} is measured during bursts, it reflects the current that circulates during every cycle of a write burst, including all the stages of the pipeline. The last access adds the time required to empty the writing pipeline and

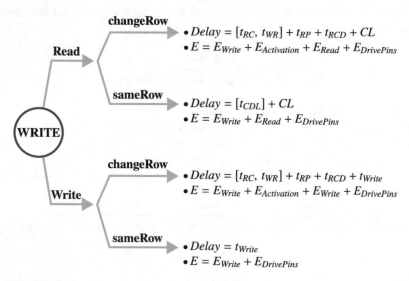

Fig. 5.14 Transitions after previous write accesses (LPSDRAM)

meet the recovery time of the last write (i.e., its propagation time). That explains the first term (E_{Write}) in the equation with P_{Write}.

$$E = (\overbrace{P_{Write} \times t_{WR}}^{\text{Finish prev. write}} + \overbrace{P_{ActPre} \times (t_{RP} + t_{RCD})}^{\text{Activation}})$$
$$+ \underbrace{P_{Read} \times CL}_{\text{Read}} + \underbrace{PDQ_r \times t_{Read}}_{\text{Drive pins}}) \times t_{CPUCycle} \tag{5.26}$$

READ After WRITE in the Active Row

When a READ command follows a WRITE, but in the same active row, the new access can proceed after meeting t_{CDL}, which represents the time needed from the last "data-in" to the next READ or WRITE command.[3] Equation (5.27) details the energy consumption for this case:

$$E = (\overbrace{P_{Write} \times t_{CDL}}^{\text{Finish write}} + \overbrace{P_{Read} \times CL}^{\text{Read}} + \overbrace{PDQ_r \times t_{Read}}^{\text{Drive pins}}) \times t_{CPUCycle} \tag{5.27}$$

WRITE After WRITE with Row Change

A WRITE command that changes the active row after a previous WRITE command has to wait for the write-recovery time (t_{WR}) and the minimum time between activations (t_{RC}). The simulator has to ensure that both timing restrictions have elapsed before continuing; both restrictions are served concurrently, not one after the other—that is why they are separated by commas and not added in Fig. 5.14. Equation (5.28) shows each of the terms that add up to the energy consumption of this case:

$$E = (\overbrace{P_{Write} \times t_{WR}}^{\text{Finish prev. write}} + \overbrace{P_{ActPre} \times (t_{RP} + t_{RCD})}^{\text{Activation}})$$
$$+ \underbrace{P_{Write} \times t_{Write}}_{\text{Write}} + \underbrace{PDQ_w \times t_{Write}}_{\text{Drive pins}}) \times t_{CPUCycle} \tag{5.28}$$

WRITE After WRITE in an Active Row

Consecutive WRITE commands to any of the active rows can proceed normally one after the other, in consecutive cycles. Energy consumption during writes is calculated using I_{DD4}, which accounts for the total current during write bursts; thus, the energy consumed by chained writes each in a different stage is appropriately reflected. Equation (5.29) shows how energy consumption is calculated in this case:

$$E = (\overbrace{P_{Write} \times t_{Write}}^{\text{Write}} + \overbrace{PDQ_w \times t_{Write}}^{\text{Drive pins}}) \times t_{CPUCycle} \tag{5.29}$$

---

[3]Normally, t_{CDL} is 1, which means that the next access can proceed in the next cycle. If its value were bigger in a device, then the following cases of WRITE-to-WRITE should also be reviewed as t_{CDL} affects both reads and writes after a write.

5.6   Simulation of LPDDR2-SDRAMs

DDR2-SDRAM devices use both edges of the clock signal to transmit data on the
bus. Figure 5.15 illustrates how LPDDR2 devices use both edges of the clock signal
to transfer data and the timing components involved in a burst of read operations. In
this section, we will explore the simulation of LPDDR2-SDRAM devices using the
specifications published by the JEDEC as reference [2]. In Chap. 6 we will extract
the concrete values for each of the parameters from the Micron datasheets [5].

The most relevant characteristic of LPDDR2-SDRAM devices is their n-prefetch
architecture, in contrast with the pipelined architecture of Mobile SDRAMs.
LPDDR2 devices are classified into S2 (2-prefetch) or S4 (4-prefetch) devices.
This means that the devices must fetch 2 or 4 words before executing an operation,
respectively. As data are transferred at both edges of the clock, the operations can be
canceled after any cycle for S2 devices, but only at every other cycle for S4 devices.
For our simulator we will consider only LPDDR2-S2 devices.

The previous consideration is relevant because the minimum burst size for DDR2
devices is 4 words, which is in accordance with a memory hierarchy model in which
DRAMs are mainly accessed to transfer cache lines. However, with the solutions
generated by our methodology, the processor may execute accesses over individual
memory words (e.g., when accessing just a pointer in a node). With an S2 device,
the memory controller can issue burst-terminate (BST) commands to limit the size
of an access to 2 words or simply chain consecutive accesses at every cycle (for
$t_{CCD} = 1$) to achieve an effective $BL = 2$. Single-word writes can be accomplished
using the write mask bits (DM) to disable writing of the odd (or even) word.

The results presented in Chap. 6 for $\mathcal{D}yn\mathcal{A}s\mathcal{T}$ solutions with LPDDR2-S2
memories incur an extra overhead for every individual access in terms of energy
consumption. Future work may study performance with S4 devices or others with a
higher prefetch index, taking into account the proportion of 1, 2 and 3-word accesses
for each concrete application.

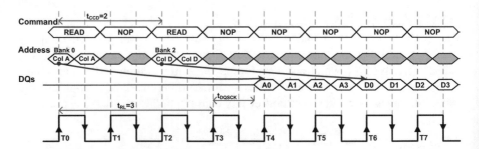

Fig. 5.15 Seamless burst read in an LPDDR2-Sx SDRAM with $t_{RL} = 3$, $BL = 4$, and $t_{CCD} = 2$.
Each read command can be issued to any bank as long as they are active. LPDDR2 devices present
two remarkable characteristics. First, data are presented on the DQ pins after $t_{RL} + t_{DQSCK}$ cycles;
t_{DQSCK}, which represents the skew between the data-pins strobe (DQS) and CLK, may be longer
than the clock period and is accounted *after* t_{RL} is over. Second, t_{CCD} may be longer than one cycle,
which delays the moment after which new commands can be issued

The working of LPDDR2 devices is considerably more complex than that of LPSDR devices. Therefore, in the following paragraphs we explain some key concepts in the simulation using excerpts directly from the JEDEC's specification.

5.6.1 Row Activations

The LPDDR2-SDRAM can accept a READ or WRITE command at time t_{RCD} after the ACTIVATE command is sent. [...] The minimum time interval between successive ACTIVATE commands to the same bank is determined by the RAS cycle time of the device (t_{RC}). [2, p. 81]

The bank(s) will be available for a subsequent row access [...] t_{RPpb} after a single-bank PRECHARGE command is issued. [2, p. 109]

We will assume again an "open-row" policy for the simulator; hence, rows remain active until an access to a different one arrives. Opening a new row requires a PRECHARGE-ACTIVATE command sequence. Subsequent READ or WRITE commands can be sent by the memory controller after t_{RCD}. Table 5.2 details the timing components that intervene in the process: t_{RPpb}, the single-bank precharging time; t_{RAS}, the row-activation time, and t_{RCD}, the ACTIVATE-to-READ-or-WRITE time.

The simulator does not implement any policy for proactively closing rows and thus does not currently account for the potential energy savings. This topic is worth future expansion.

5.6.2 Memory Reads

We will assume in the simulator that all time-consecutive read accesses form a burst; in particular, bursts can include reads to any words in the set of active rows, no matter their absolute addresses. However, LPDDR2 devices have a minimum burst size of 4 words, transmitted over two bus cycles. Smaller transfers can be achieved issuing a TERMINATE command or interrupting the current read with a new READ command:

The seamless burst read operation is supported by enabling a READ command at every other clock for $BL = 4$ operation, every 4 clocks for $BL = 8$ operation, and every 8 clocks for $BL = 16$ operation. For LPDDR2-SDRAM, this operation is allowed regardless of whether the accesses read the same or different banks as long as the banks are activated. [2, p. 91]

For LPDDR2-S2 devices, burst reads may be interrupted by other reads on any subsequent clock, provided that t_{CCD} is met. [2, p. 91]

If $t_{CCD} > 1$, single or double-word accesses will incur extra overheads.

The first access in a burst reflects the complete delay of the pipeline (t_{RL} + $t_{DQSCK} + t_{DQSQ}$), but the memory outputs one word in each subsequent bus cycle (t_{Read} is normally one cycle). In our simulator we will unify both skew terms

Table 5.2 Definition of standard working parameters for LPDDR2-SDRAMs

Name	Units	Typ.	Description
BL	32-bit words	4	Burst length (programmable to 4, 8 or 16)
t_{CK}	ns	3	Clock cycle time
t_{CCD}	Bus cycles	1	CAS-to-CAS delay
t_{DQSCK_SQ}	Bus cycles	1	DQS output access time from $CK/CK\#$ plus $DQS - DQ$ skew
t_{DQSQ}	Bus cycles	<1	$DQS - DQ$ skew
t_{DQSS}	Bus cycles	1	WRITE command to first DQS latching transition
t_{RAS}	Bus cycles	14	Row active time
t_{RCD}	Bus cycles	6	ACTIVE-to-READ-or-WRITE delay (RAS-to-CAS)
t_{RL}	Bus cycles	5	READ latency
t_{RPab}	Bus cycles	6–7	Row PRECHARGE time (all banks)
t_{RPpb}	Bus cycles	6	Row PRECHARGE time (single bank)
t_{RRD}	Bus cycles	4	ACTIVATE bank A to ACTIVATE bank B
t_{RTP}	Bus cycles	3	READ-to-PRECHARGE command delay
t_{WL}	Bus cycles	2	WRITE latency
t_{WR}	Bus cycles	5	WRITE recovery time
t_{WTR}	Bus cycles	3	WRITE-to-READ command delay
I_{DDO1}	mA	20.0	Operating one bank ACTIVE-PRECHARGE current (V_{DD1})
I_{DDO2}	mA	47.0	Operating one bank ACTIVE-PRECHARGE current (V_{DD2})
I_{DDOin}	mA	6.0	Operating one bank ACTIVE-PRECHARGE current (V_{DDca}, V_{DDq})
I_{DD3N1}	mA	1.2	Active non-power-down standby current (V_{DD1})
I_{DD3N2}	mA	23.0	Active non-power-down standby current (V_{DD2})
I_{DD3Nin}	mA	6.0	Active non-power-down standby current (V_{DDca}, V_{DDq})
I_{DD4R1}	mA	5.0	Operating burst READ current (V_{DD1})
I_{DD4R2}	mA	200.0	Operating burst READ current (V_{DD2})
I_{DD4Rin}	mA	6.0	Operating burst READ current (V_{DDca})
I_{DD4RQ}	mA	6.0	Operating burst READ current (V_{DDq})
I_{DD4W1}	mA	10.0	Operating burst WRITE current (V_{DD1})
I_{DD4W2}	mA	175.0	Operating burst WRITE current (V_{DD2})
I_{DD4Win}	mA	28.0	Operating burst WRITE current (V_{DDca}, V_{DDq})
V_{DD1}	V	1.8	Core power 1
V_{DD2}	V	1.2	Core power 2
V_{DDca}	V	1.2	Input buffer power
V_{DDq}	V	1.2	I/O buffer power
C_{IO}	pF	5.0	Input/output pins (DQ, DM, DQS_t, DQS_c) capacitance (for input, i.e., writes)
C_{LOAD}	pF	5.0	Capacitive load of the DQs (for output, i.e., reads)
t_{Read}	Bus cycles	1	
t_{Write}	Bus cycles	1	

(continued)

Table 5.2 (continued)

Name	Units	Typ.	Description
CPUToDRAMFreqFactor	n/a	4–8	*CPUFreq/DRAMFreq*
DQ	bit	32	Data pins
DQS	bit	4×2	Strobe signals for *DQ*s (differential)
DM	bit	4	Write data mask pins, one per each *DQ* byte

Timing parameters are usually provided in ns and rounded up to DRAM Bus cycles

($t_{DQSCK\_SQ} \equiv t_{DQSCK} + t_{DQSQ}$) and represent them in bus cycles (instead of ns as usually expressed in the datasheets). Figure 5.15 illustrates this situation:

> The Read Latency (t_{RL}) is defined from the rising edge of the clock on which the READ command is issued to the rising edge of the clock from which the t_{DQSCK} delay is measured. The first valid datum is available $t_{RL} \times t_{CK} + t_{DQSCK} + t_{DQSQ}$ after the rising edge of the clock where the READ command is issued. [2, p. 86]

Energy consumption is calculated for the whole duration of the access (*CL* cycles for the first one, 1 cycle for the next) because the family of $I_{DD4R\_}$ currents is measured during bursts and thus, it reflects the energy consumed by all the stages of the pipeline. A possible error source is that $I_{DD4R\_}$ might be slightly lower during the *CL* cycles of the first access as the operation is progressing through the pipeline, but there seems to be no available data in this respect.

A burst is broken by an inactivity period or an access to a different row. In those cases, the first access of the next burst will bear the complete latency. Inactivity periods (i.e., after the last word of a burst is outputted) happen if the processor accesses other memories for a while; for instance, because most of the accesses happen to an internal SRAM or cache memories are effective and thus DRAMs are seldom accessed. Accesses to words in a different row on any bank also break a burst. The simulator can be extended assuming that the PRECHARGE command was sent to the affected bank as far back in time as the last access to that bank (every access command may include an optional PRECHARGE operation) and, therefore, simulate interleaving of bank accesses with no intermediate latencies.

Contrary to the case with Mobile SDRAMs, transitions from reads to writes, and vice versa, require extra delays:

> For LPDDR2-S2 devices, reads may interrupt reads and writes may interrupt writes, provided that t_{CCD} is met. The minimum CAS-to-CAS delay is defined by t_{CCD}. [2, p. 85]
>
> The minimum time from the burst READ command to the burst WRITE command is defined by the read latency (t_{RL}) and the burst length (*BL*). Minimum READ-to-WRITE latency is $t_{RL} + \lceil t_{DQSCK_{max}}/t_{CK} \rceil + BL/2 + 1 - t_{WL}$ clock cycles. Note that if a read burst is truncated with a burst terminate (BST) command, the effective burst length of the truncated read burst should be used as *BL* to calculate the minimum READ-to-WRITE delay. [2, p. 90]

We will observe these restrictions in our simulator. We will also assume that the memory controller issues TERMINATE commands as needed for single or double-word accesses; thus, the equation is simplified because the effective *BL* is 2 and

$BL/2 = 1$. This term is kept in this simplified form in the delay and energy consumption equations presented in the rest of this section to make it explicit.

Finally, READ-to-PRECHARGE transitions have some additional restrictions in LPDDR2 devices that were not present in Mobile SDRAM devices:

> For LPDDR2-S2 devices, the minimum READ-to-PRECHARGE spacing has also to satisfy a minimum analog time from the rising clock edge that initiates the last 2-bit prefetch of a READ command. This time is called t_{RTP} (read-to-precharge). For LPDDR2-S2 devices, t_{RTP} begins $BL/2 - 1$ clock cycles after the READ command. [. . .] If the burst is truncated by a BST command or a READ command to a different bank, the effective BL shall be used to calculate when t_{RTP} begins. [2, p. 110]

5.6.3   Memory Writes

Write latencies are slightly more complex in LPDDR2 devices than in Mobile SDRAMs. The first write in a burst has an extra starting delay (t_{DQSS}). Once this delay is met, the memory controller has to provide one data word at each clock edge up to the length of the burst:

> The write latency (t_{WL}) is defined from the rising edge of the clock on which the WRITE command is issued to the rising edge of the clock from which the t_{DQSS} delay is measured. The first valid datum shall be driven $t_{WL} \times t_{CK} + t_{DQSS}$ from the rising edge of the clock from which the WRITE command is issued. [2, p. 94]

As expected, `WRITE` commands can be chained in consecutive bursts:

> The seamless burst write operation is supported by enabling a WRITE command every other clock for $BL = 4$ operation, every four clocks for $BL = 8$ operation, or every eight clocks for $BL = 16$ operation. This operation is allowed regardless of same or different banks as long as the banks are activated. [2, p. 97, Fig. 47]
>
> For LPDDR2-S2 devices, burst writes may be interrupted on any subsequent clock, provided that $t_{CCD(min)}$ is met. [2, p. 97]
>
> The effective burst length of the first write equals two times the number of clock cycles between the first write and the interrupting write. [2, p. 98, Fig. 49]

We will use this last property and the write masks (DM) to implement single-word writes in the simulator. However, this aspect may need adjustments for devices with $t_{CCD} > 1$.

As with reads, LPDDR2 devices require a minimum time to transition from writing to reading in a bank, even when no active row changes are involved:

> The minimum number of clock cycles from the burst WRITE command to the burst READ command for any bank is $[t_{WL} + 1 + BL/2 + \lceil t_{WTR}/t_{CK} \rceil]$.
>
> [. . .] t_{WTR} starts at the rising edge of the clock after the last valid input datum.
>
> [. . .] If a WRITE burst is truncated with a burst TERMINATE (BST) command, the effective burst length of the truncated write burst should be used as BL to calculate the minimum WRITE-to-READ delay. [2, p. 96, Fig. 45]

Finally, LPDDR2 devices introduce also additional restrictions for WRITE-to-PRECHARGE transitions, which were neither present for Mobile SDRAM devices:

For write cycles, a delay must be satisfied from the time of the last valid burst input data until the PRECHARGE command may be issued. This delay is known as the write recovery time (t_{WR}) referenced from the completion of the burst write to the PRECHARGE command. No PRECHARGE command to the same bank should be issued prior to the t_{WR} delay. LPDDR2-S2 devices write data to the array in prefetch pairs (prefetch $= 2$) [...]. The beginning of an internal write operation may only begin after a prefetch group has been latched completely. [...] For LPDDR2-S2 devices, minimum WRITE-to-PRECHARGE command spacing to the same bank is $t_{WL} + \lceil BL/2 \rceil + 1 + \lceil t_{WR}/t_{CK} \rceil$ clock cycles. [...] For an [sic] truncated burst, BL is the effective burst length. [2, p. 112]

Every time that the simulator has to change the active row in a bank, it must check if that bank is ready to accept new commands by evaluating if the last writing cycle plus the additional delays have already elapsed.

We calculate energy consumption during one cycle (t_{Write}) for normal write accesses; however, when a burst is finished (due to inactivity), the next command is a READ or the activity switches to a different bank, we also have to account for the extra energy consumed by the bank finishing its work in the background during t_{WR} cycles.

5.6.4 Write Data Mask

LPDDR2-SDRAM devices have a set of pins that act as byte masks during write operations. They can be used to inhibit overwriting of individual bytes inside a word. In our simulator we will assume that the memory controller exploits this capability so that individual words can be written seamlessly in one cycle (half cycle is used to access the word and the other half is wasted). In this way, the simulator does not need to distinguish internally if the written word is the even or the odd one. Instead, it can count the write access; if the next command is a write to the odd word, then it is ignored. This scheme covers the cases of writing the even, the odd or both words.

One write data mask (DM) pin for each data byte (DQ) will be supported on LPDDR2 devices, consistent with the implementation on LPDDR SDRAMs. Each data mask (DM) may mask its respective data byte (DQ) for any given cycle of the burst. Data mask has identical timings on write operations as the data bits, though used as input only, is internally loaded identically to data bits to insure matched system timing. [2, p. 103]

As with Mobile SDRAMs, the DM pins will be included in the energy consumed by the memory controller during writes.

5.6.5 Address Organization

As explained for Mobile SDRAM devices, we assume a memory address organization of "bank-row-column." Nevertheless, different addressing options should be easy to explore.

5.6.6   DRAM Refreshing

Similarly, we will ignore the effect of DRAM refreshing on energy consumption and row activations: Although LPDDR2-SDRAM devices reduce the maximum refresh period (t_{REF}) to 32 ms, its impact should still be relatively small. This point can be easily modified if needed.

5.6.7   Memory Working Parameters

We will use the parameters defined in Table 5.2 to model LPDDR2-SDRAMs. Timing parameters are usually provided by the manufacturers in ns and rounded up to DRAM bus cycles by system designers. t_{DQSCK} and t_{DQSQ} are used in ns for calculations in the JEDEC's specification; however, here we again convert them (rounding up) to bus cycles.

5.6.8   Calculations

5.6.8.1   Shortcuts

During the description of the simulation we will use the following shortcuts to simplify the writing of long equations:

$$P_{ActPre} = (I_{DDO1} \times V_{DD1} + I_{DDO2} \times V_{DD2} + I_{DDOin} \times V_{DDca}) \tag{5.30}$$

$$P_{Read} = (I_{DD4R1} \times V_{DD1} + I_{DD4R2} \times V_{DD2}$$
$$+ I_{DD4Rin} \times V_{DDca} + I_{DD4RQ} \times V_{DDq}) \tag{5.31}$$

$$P_{Write} = (I_{DD4W1} \times V_{DD1} + I_{DD4W2} \times V_{DD2} + I_{DD4Win} \times V_{DDca}) \tag{5.32}$$

$$t_{CPUCycle} = 1.0/CPUFreq \tag{5.33}$$

P_{ActPre} (5.30) is calculated using I_{DDO} in a similar way than for Mobile SDRAM devices. However, LPDDR2 devices may use two different voltage sources and thus multiple currents appear: I_{DDO1}, I_{DDO2}, and I_{DDOin} (for the input buffers). P_{ActPre} is the average power required during a series of ACTIVATE-PRECHARGE commands; thus, separating the power required for each operation is not feasible.

Similarly, LPDDR2-SDRAM devices distinguish read and write burst currents; even more, the specification separates them into their respective components. This can be seen in the equations for P_{Read} (5.31) and P_{Write} (5.32) in comparison with the case for Mobile SDRAMs.

$t_{CPUCycle}$ (5.33) is calculated using the CPU frequency defined in the platform template file and normally measured in ns.

5.6.8.2 Power of Driving Module Pins

Equations (5.34) and (5.35) give the power required to drive the module pins during
read or write accesses, respectively. The energy required for a complete $0 \rightarrow 1 \rightarrow 0$
transition corresponds to $C \times V_{DDq}{}^2$. Since DDR signals can toggle twice per clock
cycle, their effective frequency is $CPUFreq/CPUToDRAMFreqFactor$, in contrast
with the case of Mobile SDRAMs.[4]

The simulator has to calculate the power required to drive the data pins and the
data-strobe signals for reads and writes. For writes, it also needs to add the power
required to drive the write-mask pins.

$$PDQ_r = \underbrace{C_{LOAD} \times V_{DDq}{}^2}_{\text{Transition energy}} \times \underbrace{(DQ + DQS)}_{\text{Number of pins}} \frac{CPUFreq}{CPUToDRAMFreqFactor} \qquad (5.34)$$

$$PDQ_w = \underbrace{C_{IO} \times V_{DDq}{}^2}_{\text{Transition } E} \times \underbrace{(DQ + DQS + DM)}_{\text{Number of pins}} \frac{CPUFreq}{CPUToDRAMFreqFactor}$$

$$(5.35)$$

To calculate the power needed for writes, we can assume that the capacitive load
supported by the memory controller (C_{IO}) is the same than the load driven by the
memory during reads: $C_{IO} = C_{LOAD}$. Although this does not strictly correspond to
energy consumed by the memory itself, it is included as part of the energy required
to use it.

Finally, energy consumption can be calculated multiplying P_{DQ} by the length
of an access. Alternatively, the simulator could simply use $E = C \times V_{DDq}{}^2$ and
multiply by the number of transitions and the number of data pins.

5.6.8.3 Background Power

As for Mobile SDRAMs, our simulator will calculate the total energy consumed
during each memory operation. The energy consumed during standby cycles (i.e.,
cycles during which the DRAM banks were active but the DRAM was not
responding to any access) will be calculated later (5.36) using the total number of
standby cycles counted during the simulation:

$$E_{Background} = (I_{DD3N1} \times V_{DD1} + I_{DD3N2} \times V_{DD2} + I_{DD3Nin} \times V_{DDca})$$
$$\times EmptyCycles \times t_{CPUCycle}$$
$$(5.36)$$

---

[4]Once again, the amount of transitions at the DRAM pins is data-specific. Therefore, a better
approximation might be achieved by including during profiling the actual data values read or
written to the memories.

5.6.9 Simulation

We organize the simulation according to the state of the module and the type of the next operation executed. Every time that a row is activated, the simulator has to save the operation time to know when the next activation can be initiated. Similarly, the simulator needs to track the time of the last write to each bank to guarantee that the write-recovery time (t_{WR}) is met.

The following diagrams present all the timing components required during each access. Some parameters limit how soon a command can be issued, and are counted since a prior moment. Thus, they may be already elapsed at the time of the current access. These parameters are surrounded by square brackets ("[]") in the state diagrams. This applies particularly to t_{RC}, which imposes the minimum time between two ACTIVATE commands to a bank required to guarantee that the row is refreshed in the cell array after opening it.

As in the case of SDRAMs, the simulator will identify periods of inactivity for the whole module, providing the count of those longer than 1000 CPU cycles and the length of the longest one. This information may be used to explore new energy-saving techniques.

5.6.9.1   From the IDLE State

The transition from the *IDLE* state happens in the simulator's model only once, at the beginning of every DRAM module simulation (Fig. 5.16).

READ from IDLE

We will assume that the banks are precharged, so that the access has to wait for the row-activation time (t_{RCD}) and the time for the first word to appear on the data bus. Equation (5.37) gives the energy required to complete the access. E_{Read} is calculated for the whole operation time to reflect the latency of the first access in a burst. However, as data are presented on the bus for just one cycle, the energy consumed driving external pins is confined to that time (t_{Read}).

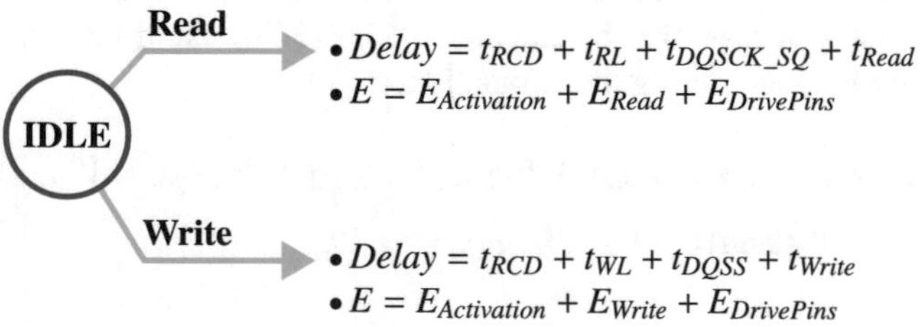

Read
- $Delay = t_{RCD} + t_{RL} + t_{DQSCK\_SQ} + t_{Read}$
- $E = E_{Activation} + E_{Read} + E_{DrivePins}$

IDLE

Write
- $Delay = t_{RCD} + t_{WL} + t_{DQSS} + t_{Write}$
- $E = E_{Activation} + E_{Write} + E_{DrivePins}$

Fig. 5.16 Initial transition for an LPDDR2-SDRAM module

$$E = E_{Activation} + E_{Read} + E_{DrivePins}$$

$$= (\overbrace{P_{ActPre} \times t_{RCD}}^{\text{Activation}} + \overbrace{P_{Read} \times (t_{RL} + t_{DQSCK\_SQ} + t_{Read})}^{\text{Read}} \tag{5.37}$$

$$+ \underbrace{PDQ_r \times t_{Read}}_{\text{Drive pins}}) \times t_{CPUCycle}$$

WRITE from IDLE

Assuming that all banks are precharged, a write access has to wait until the required row is active and then for the initial write latency ($t_{WL} + t_{DQSS}$) before the memory controller can present the first data word on the bus. Each pair of data words is then presented on the bus for one cycle (t_{Write}).

Equation (5.38) gives the energy required to complete the write. The memory controller is assumed to consume energy driving the external pins during one cycle (t_{Write}).

$$E = E_{Activation} + E_{Write} + E_{DrivePins}$$

$$= (\overbrace{P_{ActPre} \times t_{RCD}}^{\text{Activation}} + \overbrace{P_{Write} \times (t_{WL} + t_{DQSS} + t_{Write})}^{\text{Write}} \tag{5.38}$$

$$+ \underbrace{PDQ_w \times t_{Write}}_{\text{Drive pins}}) \times t_{CPUCycle}$$

5.6.9.2 From the READ State

Figure 5.17 shows the possible transitions from the READ state.

READ-to-READ with Row Change

A READ command that accesses a different row than the one currently active requires a full PRECHARGE-ACTIVATE-READ sequence. Additionally, the new access has to meet both the READ-to-PRECHARGE time (t_{RTP}) and the minimum time (t_{RAS}) that a row must be active (to ensure that the bit cells in the DRAM array are restored). Therefore, the simulator has to check the last activation and operation times for the bank and calculate the remaining part that still needs to pass—usually, it will be zero. Energy consumption is calculated with Eq. (5.39):

$$E = (\overbrace{P_{ActPre} \times (t_{RPpb} + t_{RCD})}^{\text{Activation}} + \overbrace{P_{Read} \times (t_{RL} + t_{DQSCK\_SQ} + t_{Read})}^{\text{Read}}$$

$$+ \underbrace{PDQ_r \times t_{Read}}_{\text{Drive pins}}) \times t_{CPUCycle} \tag{5.39}$$

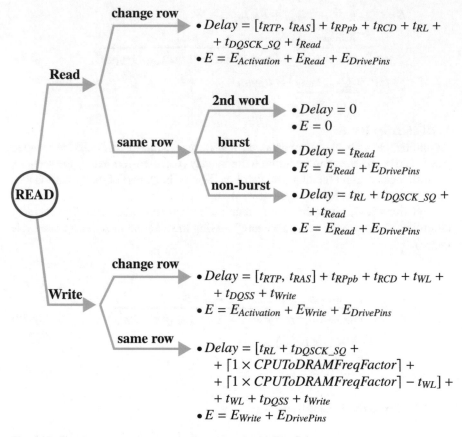

Fig. 5.17 Transitions after previous read accesses (LPDDR2-SDRAM)

READ-to-READ in the Active Row

Consecutive reads present three possibilities. The first one is that the access corresponds to the second word of a two-word transfer: The LPDDR2 module transfers two words per bus cycle, but the simulator sees the individual accesses from the memory access trace. The simulator can tackle with this situation by accounting for the total delay and energy during the first access; if the immediately consecutive access corresponds to the next word, both the latency and the energy consumption are counted as zero.

The second possibility is that the access belongs to an ongoing burst or to the first access in a chained ("seamless") burst. As the time to fill the pipeline was already accounted for the first access in the burst, the delay of this access with respect to the previous is only one cycle. Figure 5.15 presented this case: The first READ command with $BL = 4$ at T_0 was answered during T_4 and T_5, and the next READ (originated at T_2) was answered during T_6 and T_7. To achieve $BL = 2$ accesses in $t_{CCD} = 1$ devices, READ commands have to be presented in consecutive cycles, thus effectively terminating the previous burst ($BL = 4$ is the minimum supported by the standard, but for LPDDR2-S2 devices with $t_{CCD} = 1$ this mode of operation is allowed).

Equation (5.40) accounts for the energy consumption of this case. This access corresponds to reading and transferring two words; thus, the simulator has to save the address to check if the next access corresponds to the word transferred in the second half of the bus cycle (i.e., the previous paragraph).

$$E = (\overbrace{P_{Read} \times t_{Read}}^{\text{Read}} + \overbrace{PDQ_r \times t_{Read}}^{\text{Drive pins}}) \times t_{CPUCycle} \tag{5.40}$$

Finally, if the access starts a new burst that is not consecutive (in time) to the previous one, it bears the full starting cost for the burst. Equation (5.41) details the energy consumption for this case:

$$E = (\overbrace{P_{Read} \times (t_{RL} + t_{DQSCK\_SQ} + t_{Read})}^{\text{Read}} + \overbrace{PDQ_r \times t_{Read}}^{\text{Drive pins}}) \times t_{CPUCycle} \tag{5.41}$$

READ-to-WRITE with Row Change

A WRITE command that accesses a row that is not the currently active one starts a full PRECHARGE-ACTIVATE-WRITE cycle. Again, the new access has to meet both the READ-to-PRECHARGE time (t_{RTP}) and the minimum time (t_{RAS}) that a row must be active. Therefore, the simulator has to check the last activation and operation times for the bank and calculate the remaining part that still needs to pass. The memory controller can present the first data word on the data bus once the new row is active ($t_{RPpb} + t_{RCD}$) and the starting delays for the write burst are met ($t_{WL} + t_{DQSS}$). Each pair of words is presented on the bus during one cycle (t_{Write}).

Equation (5.42) shows the details of energy consumption. The energy consumed during the first cycles of a burst is accounted for this access.

$$E = (\overbrace{P_{ActPre} \times (t_{RPpb} + t_{RCD})}^{\text{Activation}} + \overbrace{P_{Write} \times (t_{WL} + t_{DQSS} + t_{Write})}^{\text{Write}}$$
$$+ \underbrace{PDQ_w \times t_{Write}}_{\text{Drive pins}}) \times t_{CPUCycle} \tag{5.42}$$

READ-to-WRITE in the Active Row

A WRITE command can follow (or interrupt) a previous READ command to the active row after a minimum delay calculated as shown in Fig. 5.17. As explained previously, we assume that the minimum burst size for LPDDR2-S2 devices is in effect $BL = 2$, so that the factor $BL/2$ in the specification is simplified to 1 in the figure. However, all the terms (and rounding operators) are kept to make explicit their existence—the simulator multiplies internally all timings by *CPUToDRAMFreqFactor*. After that, the write operation has the usual starting delay.

Equation (5.43) shows the details of energy consumption:

$$E = (\overbrace{P_{Write} \times (t_{WL} + t_{DQSS} + t_{Write})}^{\text{Write}} + \overbrace{PDQ_w \times t_{Write}}^{\text{Drive pins}}) \times t_{CPUCycle} \qquad (5.43)$$

5.6.9.3   From the WRITE State

Figure 5.18 shows the possible transitions from the WRITE state.

WRITE-to-READ with Row Change

A READ command that requires a row-change in a bank can follow (or interrupt) a previous writing burst after a minimum delay (Fig. 5.18), which includes the

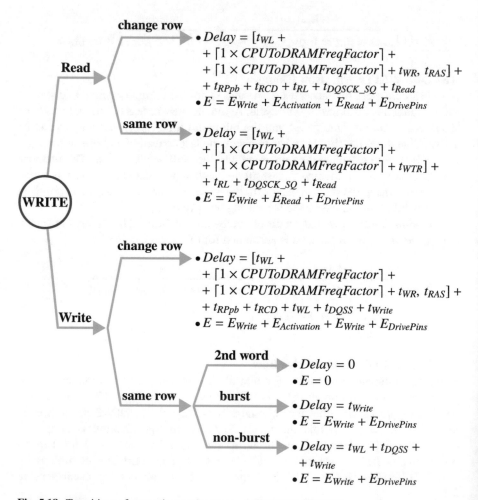

Fig. 5.18 Transitions after previous write accesses (LPDDR2-SDRAM)

write-recovery time (t_{WR}). Additionally, the simulator must check that t_{RAS} has already elapsed since the last ACTIVATE command to that bank. Both conditions are independent and thus appear separated by commas in the diagram. After that, the normal ACTIVATE-to-READ delays apply.

Equation (5.44) shows each of the terms that add up to the energy consumption of this case, including the energy consumed during the final phase of the previous write operation (E_{Write}):

$$E = (\overbrace{P_{Write} \times t_{WR}}^{\text{Finish prev. write}} + \overbrace{P_{ActPre} \times (t_{RPpb} + t_{RCD})}^{\text{Activation}}$$
$$+ \overbrace{P_{Read} \times (t_{RL} + t_{DQSCK\_SQ} + t_{Read})}^{\text{Read}} + \overbrace{PDQ_r \times t_{Read}}^{\text{Drive pins}}) \times t_{CPUCycle}$$
(5.44)

WRITE-to-READ in the Active Row

As in the previous case, A READ command in the active row of a bank can follow (or interrupt) a previous writing burst after a minimum delay (Fig. 5.18). In this case, t_{WR} is substituted by t_{WTR}, the WRITE-to-READ command delay. However, as a change of active row is not involved in this case, the normal delay for a READ command applies directly after that.

Equation (5.45) details the energy consumption for this case, including the energy consumed during the final phase of the previous write operation (E_{Write}):

$$E = (\overbrace{P_{Write} \times t_{WTR}}^{\text{Finish write}} + \overbrace{P_{Read} \times (t_{RL} + t_{DQSCK\_SQ} + t_{Read})}^{\text{Read}}$$
$$+ \underbrace{PDQ_r \times t_{Read}}_{\text{Drive pins}}) \times t_{CPUCycle}$$
(5.45)

WRITE-to-WRITE with Row Change

A WRITE command to a different row can follow (or interrupt) a previous writing burst to the same bank after a minimum delay (Fig. 5.18), which includes the write-recovery time (t_{WR}). Additionally, the simulator needs to check that t_{RAS} has already elapsed (in parallel, not consecutively) since the last ACTIVATE command to that bank. After that, the normal ACTIVATE-to-WRITE delays apply.

Equation (5.46) shows each of the terms that add up to the energy consumption of this case, including the energy consumed during the final phase of the previous write operation (E_{Write}):

$$E = (\overbrace{P_{Write} \times t_{WR}}^{\text{Finish prev. write}} + \overbrace{P_{ActPre} \times (t_{RPpb} + t_{RCD})}^{\text{Activation}}$$
$$+ \underbrace{P_{Write} \times (t_{WL} + t_{DQSS} + t_{Write})}_{\text{Write}} + \underbrace{PDQ_w \times t_{Write}}_{\text{Drive pins}}) \times t_{CPUCycle}$$
(5.46)

WRITE-to-WRITE in the Active Row

Consecutive WRITE commands to the active row of a bank come in three flavors. First, if the access corresponds to the second word of a DDR two-word transfer, both the latency and the energy consumption are counted as zero because the simulator already accounted for them when it encountered the first access.

Second, if the access belongs to an ongoing burst or is the first access in a chained ("seamless") burst, the delay of this access with respect to the previous is only one cycle. LPDDR2-S2 devices support $BL = 2$ bursts if a WRITE command follows immediately the previous, provided that the device supports $t_{CCD} = 1$.

Equation (5.47) accounts for the energy consumption of this case. This access corresponds to the first of two words; thus, the simulator saves the address to check if the next access corresponds to the word transferred in the second half of the bus cycle (i.e., as per the previous paragraph).

$$E = (\overbrace{P_{Write} \times t_{Write}}^{\text{Write}} + \overbrace{PDQ_w \times t_{Write}}^{\text{Drive pins}}) \times t_{CPUCycle} \qquad (5.47)$$

Finally, if the access starts a new burst that is not consecutive (in time) to the previous one, it bears the full starting cost for the burst. Contrary to the case of Mobile SDRAMs, in LPDDR2-DRAMs this first WRITE access in a burst has an extra delay of $t_{WL} + t_{DQSS}$. Equation (5.48) details the energy consumption for this case:

$$E = (\overbrace{P_{Write} \times (t_{WL} + t_{DQSS} + t_{Write})}^{\text{Write}} + \overbrace{PDQ_w \times t_{Write}}^{\text{Drive pins}}) \times t_{CPUCycle} \qquad (5.48)$$

5.6.10 A Final Consideration

The simulation of LPDDR2 devices is a complex topic that depends on several factors that are difficult to measure, such as the currents that circulate through the device during the initial cycles of a read burst. Additionally, although a great deal of care has been put into the development of the equations and diagrams presented in this chapter, the reader might find errors or inaccuracies in them. Any mistakes that affected access costs might change the relative performance of cache or SRAM-based solutions as the first ones tend to execute many more accesses over the DRAM. For example, if the cost of the operations were found to be lower than as calculated by our simulator, the distance between cache and SRAM solutions would narrow. Correspondingly, if the real cost were found to be higher, the distance would increase.

A form of error-bounding can be found in the ratio between the number of DRAM accesses performed by each solution. For example, if an SRAM solution produces 0.75 times the number of DRAM accesses than a cache solution, the difference in energy consumption between them should be bounded by approximately that same factor, save differences in the number of bank activations produced by each one. This difference is increased with the factor between SRAM-accesses multiplied by their energy consumption and cache-accesses multiplied by their respective cost, the latter being usually higher for the same capacity.

References

1. Dodd, J.M.: Adaptive page management. US Patent 7,076,617 B2. Intel Corporation, 2006
2. JEDEC: Low Power Double Data Rate 2 (LPDDR2)—JESD209-2E. JEDEC Solid State Technology Association, Arlington (2011)
3. Marchal, P., Gómez, J.I., Piñuel, L., Bruni, D., Benini, L., Catthoor, F., Corporaal, H.: SDRAM-energy-aware memory allocation for dynamic multi-media applications on multi-processor platforms. In: Proceedings of Design, Automation and Test in Europe (DATE) (2003)
4. MICRON: Mobile LPSDR SDRAM—MT48H32M32LF/LG Rev. D 1/11 EN. Micron Technology, Boise (2010)
5. MICRON: Mobile LPDDR2 SDRAM—MT42L64M32D1 Rev. N 3/12 EN. Micron Technology, Boise (2012)
6. Zhang, Z., Zhu, Z., Zhang, X.: A permutation-based page interleaving scheme to reduce row-buffer conflicts and exploit data locality. In: Proceedings of the Annual ACM/IEEE International Symposium on Microarchitecture (MICRO), pp. 32–41. ACM Press, Monterey (2000). https://doi.org/10.1145/360128.360134

Chapter 6
Experiments on Data Placement: Results and Discussion

In this chapter we will apply our methodology on three case studies to assess the improvements that can be attained. We will evaluate more than a hundred different platforms for each case, including multiple combinations of SRAMs and caches in conjunction with two technologies of main memory (Mobile SDRAM and LPDDR2-SDRAM). The first case study will take us on a detailed journey through all the phases of the optimization process with a model of a network of wireless sensors where the devices have to process a moderate amount of data; the focus here is on reducing energy consumption to prolong battery life. The other two cases will deepen in the analysis of the improvements obtained for different types of applications, highlighting the effects of data placement with explicitly addressable memories in comparison with traditional cache memories. Specifically, the second experiment uses the core of a network routing application as an example of high performance data processing, whereas the third experiment is a small DDT-intensive benchmark. These experiments compare the cost of executing an application in a platform with hardware-based caches or in a platform with one (or several) explicitly addressable on-chip SRAMs managed via the dynamic memory manager as proposed in this text.

Although we will explore the performance of the different case studies on a myriad of different platforms, system designers do not need to do the same. In fact, if the platform is defined, they just need to apply once our methodology on their application to obtain the final data placement. Here, we do the additional experiments to generate a wide comparison base.

6.1 Methodology Implementation in the DynAsT Tool

The methodology has been implemented as a functional tool, $\mathcal{D}yn\mathcal{A}s\mathcal{T}$, which can be used to improve the placement of dynamic data structures on the memory

© Springer Nature Switzerland AG 2020   143
M. Peón Quirós et al., *Heterogeneous Memory Organizations*
in Embedded Systems, https://doi.org/10.1007/978-3-030-37432-7_6

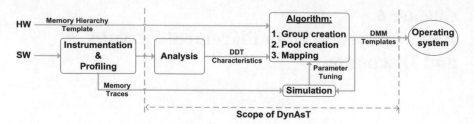

Fig. 6.1 Methodology implementation in $\mathcal{D}yn\mathcal{A}s\mathcal{T}$

subsystem of an existing platform, or to steer the design of a new platform according to the particular needs of the future applications. $\mathcal{D}yn\mathcal{A}s\mathcal{T}$ performs the analysis of the information obtained during profiling, the grouping and mapping steps, and the optional simulation (Fig. 6.1). We use it through this chapter to show the results obtained in several examples with a concrete implementation of the methodology.

The first action of the tool is to analyze the traces obtained during profiling to infer the characteristics of each DDT. During the grouping step, which is implemented using several heuristics to limit its complexity, it analyzes the footprint of those DDTs that have a similar amount of accesses and tries to cluster them, matching the "valleys" in the memory footprint of some with the "peaks" of others. Pool formation is a stub to introduce any dynamic memory management techniques currently available. Finally, $\mathcal{D}yn\mathcal{A}s\mathcal{T}$ produces a mapping of every pool, and hence of the dynamic data objects that it will contain, over the memory modules of the target platform. For this task, the tool considers the characteristics of each group/pool and the properties of each memory module in the platform. The mapping step is simpler because the pools can be split over several memory resources even if their address ranges are not consecutive. The different steps interact with each other through data abstractions (e.g., DDT behavior, group, group behavior, pool); hence, they can be improved independently. For example, further research may devise new grouping algorithms or the mapping step could be integrated with the run-time loader of the operating system.

As an optional feature, $\mathcal{D}yn\mathcal{A}s\mathcal{T}$ also includes the memory hierarchy simulator prescribed in the methodology to evaluate the properties of the generated placement solutions. The simulation results can be used to iterate over the grouping, pool formation, and mapping steps to tune the placement (e.g., changing the values of the parameters in each algorithm).

The output of the tool is a description of the dynamic memory managers that can be used to implement the solution at run-time. This description includes their internal algorithms, the DDTs that will be allocated into each of them, their size, and their position in the memory address space (Fig. 6.2).

The data placement generated by $\mathcal{D}yn\mathcal{A}s\mathcal{T}$ is implemented as follows. During instrumentation, each DDT is assigned a unique numerical ID. The grouping phase of the tool decides which DDTs share a pool: The correspondence between groups and pools is direct. During application execution, the DMM receives not only the

Fig. 6.2 The outcome of the methodology is the description of the pools, the list of DDTs that will be allocated in each one, and their placement on memory resources

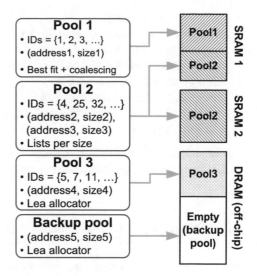

size of the request, but also the type of the object being created—that is, the ID assigned to its DDT. With that extended information, the DMM identifies the pool that corresponds to that ID and calls the (sub)DMM that manages that pool. If the local DMM does not have enough resources left in its heaps, the request is redirected to the backup pool. The backup pool resides in the last eligible memory resource— it is important that it remains seldom used as accesses to any objects allocated there will incur the highest cost—and receives allocations that exceed the maximum footprint values measured during profiling for each pool. The backup pool is not the same than pools that are explicitly mapped to a DRAM, something that happens because they are the least accessed or (in future work) because their access pattern makes them the most appropriate for the characteristics of a DRAM and thus get less priority for the use of the available SRAMs.

In summary, $DynAsT$ currently implements a set of simple algorithms and heuristics that can be independently improved in the future. Despite its simplicity, it shows the promising results that can be attained with a data placement that takes into account the characteristics of the application data types.

6.1.1   Parameters Used in the Experiments

The interested reader should be able to reproduce the results presented here either with the same simulator or building an equivalent one based on the descriptions of the previous chapters. Throughout all the experiments of this chapter, we used the following methodology parameters unless otherwise specified:

- $MaxGroups = +\infty$;
- $MinIncFPB = 1.0$ and $MinIncExpRatio = 1.0$: Any increase on FPB or exploitation ratio is accepted;
- $SpreadPoolsInDRAMBanks = True$;
- $MinMemoryLeftOver = 8\,B$;
- $PoolSizeIncreaseFactor = 1.0$;
- $PoolSizeIncreaseFactorForSplitPools = 1.3$;
- $UseBackupPool = True$.

Further improvements might be achieved with a careful tuning of these parameters to the characteristics of each particular application.

6.2 Description of the Memory Hierarchies

Here we present the technical parameters of the different types of memories used in our experiments. The technical parameters of the SRAM and cache memories were obtained via Cacti [2] using a 32 nm feature size. Tables 6.1 and 6.2 detail their respective technical parameters. Interestingly, the values for energy consumption in Table 6.2 present some unexpected variations. For example, the energy consumed during an access by a 4 KB direct mapped cache is higher than the energy consumed by a 32 KB 2-way associative cache. We assume that these values are correct and correspond to the particular characteristics of signal routing inside big circuits. The following quote by Hennessy and Patterson [7, p. 77, Fig. 2.3] points in the same direction regarding the evolution of access times:

> Access times generally increase as cache size and associativity are increased. [...] The assumptions about cache layout and the complex trade-offs between interconnect delays (that depend on the size of a cache block being accessed) and the cost of tag checks and multiplexing lead to results that are occasionally surprising, such as the lower access time for a 64 KB with two-way set associativity versus direct mapping. Similarly, the results with eight-way set associativity generate unusual behavior as cache size is increased. Since

Table 6.1 Technical parameters of the (on-chip) SRAMs

Size	Energy (nJ)	Latency (cycles)	Area (mm^2)
512 B	<0.001	1	0.003
1 KB	0.001	1	0.005
4 KB	0.002	1	0.012
16 KB	0.004	1	0.045
32 KB	0.005	1	0.112
64 KB	0.007	1	0.185
256 KB	0.013	2	0.781
512 KB	0.025	2	1.586
1 MB	0.028	4	2.829
4 MB	0.077	6	11.029

Table 6.2 Technical parameters of the cache memories

Size	Associativity	Line size (words)	Energy (nJ)	Latency (cycles)	Area (mm²)
4 KB	Direct	16	0.154	1	0.021
32 KB	2 ways	16	0.102	1	0.075
64 KB	4 ways	16	0.119	1	0.140
16 KB	16 ways	16	0.166	1	0.137
32 KB	16 ways	16	0.166	1	0.203
32 KB	16 ways	4	0.024	1	0.100
64 KB	16 ways	16	0.179	1	0.263
64 KB	16 ways	4	0.030	1	0.158
256 KB	16 ways	16	0.250	2	0.609
256 KB	16 ways	4	0.068	2	0.533
512 KB	16 ways	16	0.345	2	1.069
1 MB	16 ways	16	0.509	4	2.088
4 MB	16 ways	16	2.124	6	8.642

such observations are highly dependent on technology and detailed design assumptions, tools such as CACTI serve to reduce the search space rather than precision analysis of the trade-offs.

We configured the cache memories for the experiments using as reference the ARM Cortex-A15 [1]: a line size of 64 bytes (16 words), associativity of up to 16 ways, capacity of up to 4 MB (for the L2 caches), and least recently used (LRU) replacement policy. Here, we present multiple different configurations to explore the implications of these design options. Among others, we use configurations with sizes varying from 4 KB to 4 MB, with associativity degrees ranging from direct mapped ("D"), to 2-ways ("A2"), 4-ways ("A4"), and 16-ways ("A16"); and with either 16 (the default) or 4 ("W4") words per line. All the cache memories use an LRU replacement policy. Finally, we also include in all the experiments special configurations labeled as "lower bound" that represent the minimum theoretical cost for a cache memory with a big size (256 MB), but a small cost (comparable to that of a 4 KB one)—thus providing a minimum bounds for execution time and energy consumption.

With our methodology, we generate solutions for configurations that consist of one or several on-chip SRAMs and an external DRAM. These configurations are labeled as "SRAM," where their labels enumerate the independent memory modules that are included. For instance, a configuration labeled as "SRAM 8 × 512 KB" has 8 modules of 512 KB each. All the experiments include a base configuration with SRAM modules of 512 B, 1 KB, 32 KB, and 256 KB to serve as a common reference. Finally, a configuration labeled as "lower bound" with an on-chip SRAM of 1 GB and the properties of a 4 KB memory is also included to provide a lower cost bound. Figures 6.3 and 6.4 show how the different memories are positioned in terms of energy consumption per access and area, respectively. The much lower

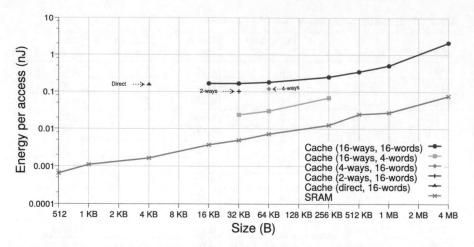

Fig. 6.3 Energy per access for the cache and SRAM configurations used in the experiments (logarithmic scale). Remarkably, SRAMs are much more energy efficient than caches of equivalent capacity, which motivates our pursue of efficient data placement techniques

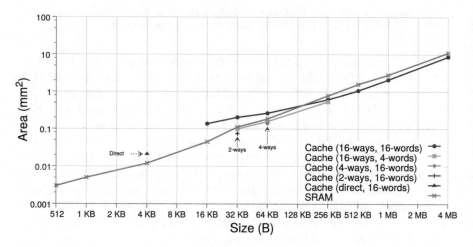

Fig. 6.4 Area of the different cache and SRAM configurations used in the experiments (logarithmic scale). The counterintuitive bigger size of the SRAMs of larger capacity with respect to equivalent caches may obey to the particularities of the tools used for modeling

energy consumption observed in Fig. 6.3 for SRAMs compared to caches of the same capacity explains that, if the placement decisions are correctly taken, these solutions achieve more energy-efficient operating points.

As we have seen previously, the methodology allows the designer to easily combine SRAMs with caches. Although in this set of experiments the effect of these combinations is not relevant, they may become useful in cases where a small set of data instances gets a low but still significant number of accesses (with some locality) in the DRAM modules, or if the instances of a DDT receive very different numbers of access along execution time, as may be the case with balanced trees [3, Chap. 7].

In respect to the DRAMs, the Mobile SDRAM and LPDDR2-SDRAM modules are configured according to the technical parameters that we have seen in Sects. 5.5 and 5.6.

6.3   Case Study 1: Wireless Sensors Network

In this first case study we will obtain a global view of the process by applying the methodology step by step to optimizing an application for several platform configurations with SRAMs of varying sizes. The subject application is a model of a network of wireless sensors that can be distributed over wide areas. Each sensor extracts information of its surroundings and sends it through a low-power radio link to the next sensor in the path towards the base station. The networks constructed in this way are very resilient to failures and terrain irregularities because the nodes can find new paths. They may be used for applications such as weather monitoring [8] and fire detection in wild areas or radio signal detection in military operations. Each sensor keeps several hash tables and input and output queues to provide some basic buffering capabilities. The sensors use dynamic memory due to the impossibility of determining the network dependencies and the need of adjusting the use of resources to what is strictly needed at each moment. As a consequence, the sizes of the hash tables and network queues, among others, need to be scaled. The application model creates a network of sensors and monitors one in the middle of the transmission chain; that node forwards the information from more distant ones and gathers and sends its own information periodically. In the next paragraphs we will see how the different steps of the methodology are performed and their outcome.

6.3.1   Profiling

Instrumenting the application as explained in Sect. 4.6 requires modifying 9 lines out of 3484. That means that just a 0.26% of the source code has to be modified both for profiling and deployment. After instrumentation, we ran the model normally during 4 h of real time, producing a log file of 464 MB.

6.3.2   Analysis

The analysis of the log file with the tool that implements our methodology, $\mathcal{D}yn\mathcal{A}s\mathcal{T}$, takes about 29 s, and identifies 21 distinct DDTs. The frequency of accesses per byte (FPB) of each DDT ranges from 8.8×10^5 accesses per byte down to 0.61 accesses per byte. The maximum footprint required for a given DDT is 24,624 B and the minimum, 12 B. As expected, the size of the internal dynamic data structures of the application varies along time. For example, the hash table for the active neighbors uses internal arrays of 804 B, 1644 B, 3324 B, 6684 B, and 13,404 B. $\mathcal{DynAsT}$ detects this and separates the different instances of the hash table DDT (classifying them according to their different sizes) for the next steps.

6.3.3   Grouping

The grouping step runs in about 70 s with the parameters explained at the beginning of this section. The output is a total of 12 groups containing the initial set of 21 DDTs. One of the groups has 5 DDTs, one group gets 3 DDTs, three groups hold 2 DDTs, and the last seven groups contain just 1 DDT. The grouping step manages to reduce the total footprint of the application (compared to the case of one DDT per pool) from 110,980 B to 90,868 B, thus achieving an overall footprint reduction of an 18.12% and reducing from 21 to 12 the number of pools that have to be managed (−42.9%). For the five groups that combine several DDTs, the respective memory footprint reductions are 32.6%, 56.4%, 19.9%, 51.7%, and 19.6% when compared to the space that would be required if each DDT were mapped into an independent pool.

We can obtain more insights into the grouping step looking at Fig. 6.5. DDT_2 and DDT_6 are identified as compatible and combined in the same group, yielding a significant reduction in memory usage. If each DDT were mapped into an independent pool, that is, building per-DDT pools, the total memory space required would be 1032 B (represented by a thick black horizontal line). With grouping, the required footprint is reduced to just 696 B (green horizontal line), a reduction of 32.6%.

The figure also shows how the maximum memory footprint of the DDTs and the group is determined by instantaneous peaks in the graph. The designer might use the `PoolSizeIncreaseRatio` parameter during the mapping phase to adjust the final footprint so that it covers only the space required during most of the time, relying on the use of a backup pool in DRAM to absorb the peaks. However, there is an important drawback to this: Although the freed space could then be exploited by other DDTs (instead of staying unused when the footprint of the group is lower), it is possible that the instances created during the peak footprint periods, and that would have to be allocated in the backup pool, get so many accesses that the total system performance is reduced. After all, the instances of the DDTs included in the group are expected to have a higher FPB than the instances of DDTs included in the next groups. This consideration shows the importance of the grouping step to improve the exploitation ratio of the memory space assigned to each DDT and of the simulation step to predict the consequences of different parameter values.

Finally, the evolution of the footprint of the two DDTs and the combined group can be observed in more detail in the inset in Fig. 6.5, which presents a randomly chosen period during the execution time of the application. The fact that the footprint peaks of DDT_2 are not coincident with the peaks of DDT_6 is the factor that enables their grouping.

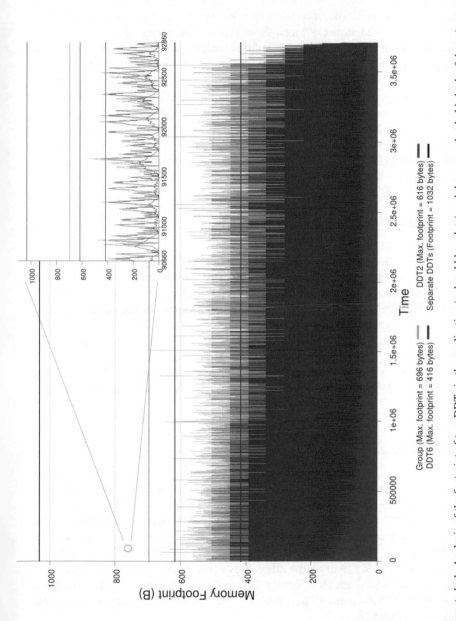

Fig. 6.5 Case study 1: Analysis of the footprint of two DDTs in the application (red and blue plots) and the group that holds both of them (green plot). The combined footprint of the group (green horizontal bar) is significantly lower than the added footprint that would be needed if the DDTs were assigned to independent pools (black horizontal bar). Inset: Zoom over a randomly chosen area of 2200 (allocation time) instants

6.3.4 Pool Formation

For this experiment, we select an allocator that always performs coalesce and split and that uses one free list per allocation size. When a block is allocated, the leftover space in the assigned memory block is separated and inserted into a list of free blocks; when a block is deallocated, it is fused with the previous and/or next blocks if they are also free before being inserted into the matching list of free blocks.

As we saw in Sect. 5.1.2, our simulator does not consider the accesses of the dynamic memory managers themselves (only the application accesses are traced). Therefore, it cannot be used to analyze the impact of the different DMMs generated during this phase. Nevertheless, introducing this functionality is straightforward by simply linking the code of the DMMs with the simulator itself so that every time that the DMM code needs to read or update an internal structure, these accesses are reproduced in the simulated platform.

6.3.5 Mapping

Mapping is the first platform-dependent step in the methodology. In this case, we run the mapping and simulation steps for more than 200 different platforms to verify that the methodology achieves better results than cache memories for many different potential platform configurations. Due to the small footprint of this application, we have included in these experiments several configurations with reduced memory subsystems (down to just 32 KB). However, it is important to stress that, if the platform is already decided, the designer has to run the mapping and simulation steps just once with the platform parameters. The execution of this step with the parameters detailed at the beginning of this section requires less than 1 s per configuration.

6.3.6 Simulation

The simulator evaluates the performance of the mapping solution for every platform using the memory trace obtained during profiling. Each simulation requires about 16 s (the simulation process is considerably faster than the analysis). As the DRAM modules are seldom accessed in most platforms, we can assume that the memory controller can drive the chips into one of the power-saving modes and thus configure the simulator to discard their active-idle energy consumption. Table 6.3 presents the results, normalized taking as reference the platforms with only DRAM modules (that is, without cache memories or SRAMs), which in this experiment correspond to platforms 00 and 17.

The experiments show that even for small sizes, cache memories improve significantly the performance of the system. However, the solutions obtained applying our methodology achieve even bigger gains. To get a better measure of

Table 6.3 Case study 1: Performance of the solutions obtained with our methodology versus cache-based solutions

Platform	Energy mJ	Energy %	Time (×10^6) cycles	Time %	Page misses %	DRAM accesses %	Total accesses %
(Mobile SDRAM)							
00. Only DRAM	360.01	100.0	2 132.5	100.0	100.0	100.0	100.0
01. Cache: L1 = 256 KiB (A16)	21.88	6.1	174.5	8.2	<0.1	<0.1	100.1
02. Cache: L1 = 256 KiB (A16,W4)	6.03	1.7	174.5	8.2	<0.1	<0.1	100.1
03. Cache: L1 = 16 KiB (A16), L2 = 256 KiB (A16)	51.47	14.3	267.6	12.5	<0.1	<0.1	300.5
04. Cache: L1 = 32 KiB (A16), L2 = 256 KiB (A16)	33.29	9.2	178.8	8.4	<0.1	<0.1	201.8
05. SRAM: 512 B, 1 KiB, 32 KiB, 256 KiB	0.24	0.1	90.1	4.2	0	0	100.0
06. Cache: L1 = 64 KiB (A16)	15.94	4.4	88.3	4.1	<0.1	0.1	100.2
07. Cache: L1 = 64 KiB (A16,W4)	2.94	0.8	88.8	4.2	0.1	0.1	100.2
08. SRAM: 512 B, 1 KiB, 64 KiB	0.46	0.1	87.7	4.1	<0.1	0.1	100.0
09. SRAM: 64 KiB	0.93	0.3	88.1	4.1	<0.1	0.1	100.0
10. Cache: L1 = 32 KiB (A16)	106.71	29.6	434.3	20.4	9.7	35.0	168.7
11. Cache: L1 = 32 KiB (A16,W4)	36.08	10.0	275.4	12.9	11.5	9.0	118.0
12. SRAM: 512 B, 1 KiB, 32 KiB	10.62	2.9	132.9	6.2	2.2	3.4	100.0
13. SRAM: 512 B, 1 KiB, 16 KiB, 32 KiB	6.39	1.8	112.3	5.3	1.0	2.1	100.0
14. SRAM: 32 KiB	12.23	3.4	141.1	6.6	2.9	3.7	100.0
15. SRAM: LowerBound	0.14	<0.1	87.1	4.1	0	0	100.0
16. Cache: LowerBound(D)	13.52	3.8	87.3	4.1	<0.1	<0.1	100.1

(continued)

Table 6.3 (continued)

Platform	Energy mJ	Energy %	Time ($\times 10^6$) cycles	Time %	Page misses %	DRAM accesses %	Total accesses %
(LPDDR2-SDRAM)							
17. Only DRAM	229.57	100.0	1 315.6	100.0	100.0	100.0	100.0
18. Cache: L1 = 256 KiB (A16)	21.84	9.5	174.4	13.3	<0.1	<0.1	100.1
19. Cache: L1 = 256 KiB (A16,W4)	6.01	2.6	174.5	13.3	<0.1	<0.1	100.1
20. Cache: L1 = 16 KiB (A16), L2 = 256 KiB (A16)	54.64	23.8	283.1	21.5	<0.1	<0.1	318.0
21. Cache: L1 = 32 KiB (A16), L2 = 256 KiB (A16)	32.80	14.3	176.5	13.4	<0.1	<0.1	199.4
22. SRAM: 512 B, 1 KiB, 32 KiB, 256 KiB	0.24	0.1	90.1	6.8	0	0	100.0
23. Cache: L1 = 64 KiB (A16)	16.21	7.1	90.0	6.8	0.5	0.6	101.1
24. Cache: L1 = 64 KiB (A16,W4)	2.89	1.3	89.1	6.8	0.4	0.1	100.2
25. SRAM: 512 B, 1 KiB, 64 KiB	0.37	0.2	87.4	6.6	0.1	0.1	100.0
26. SRAM: 64 KiB	0.83	0.4	88.0	6.7	0.1	0.1	100.0
27. Cache: L1 = 32 KiB (A16)	49.30	21.5	258.5	19.6	26.9	34.3	167.3
28. Cache: L1 = 32 KiB (A16,W4)	34.85	15.2	323.6	24.6	47.6	15.8	129.7
29. SRAM: 512 B, 1 KiB, 32 KiB	7.35	3.2	117.6	8.9	0.3	3.4	100.0
30. SRAM: 512 B, 1 KiB, 16 KiB, 32 KiB	4.65	2.0	106.0	8.1	0.1	2.1	100.0
31. SRAM: 32 KiB	8.85	3.9	126.8	9.6	3.2	3.7	100.0
32. SRAM: LowerBound	0.14	0.1	87.1	6.6	0	0	100.0
33. Cache: LowerBound(D)	13.49	5.9	87.2	6.6	<0.1	<0.1	100.1

The entries in the first half of the table, which correspond to platforms with a Mobile SDRAM, are normalized using as reference platform 00. Correspondingly, the entries in the second part (platforms with an LPDDR2) are normalized taking as reference platform 17

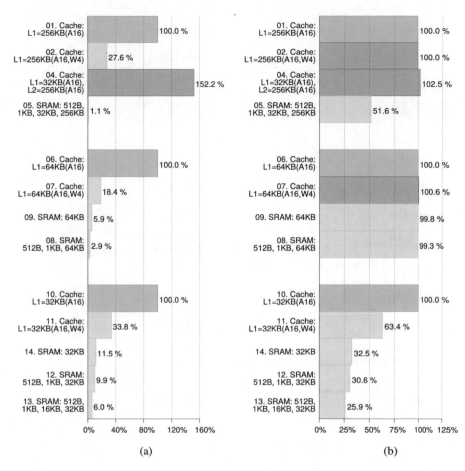

Fig. 6.6 Case study 1: Comparison of SRAM-based solutions with solutions based on caches of equivalent capacity for platforms with Mobile SDRAM. Each group of results is normalized to the uppermost bar (in blue), which represents the performance for a cache of a given size. Red bars mark configurations that have worst performance than their reference. (**a**) Energy consumption. (**b**) Cycles memory subsystem

this improvement, Figs. 6.6 and 6.7 show a direct comparison between both types of solutions for various memory subsystem sizes. Each of these figures shows three different comparisons, where the results of each group are normalized to the value of the first bar in the group (in blue).

For instance, in Table 6.3 we see that platform "06. Cache: L1 = 64 KiB (A16)" reduces the energy consumption in comparison with the platform that has only DRAM, platform 00, down to a 4.4%. In Fig. 6.6a, this is taken as the base case for the second group of bars. There, we see that modifying the length of the cache lines (platform "07. Cache: L1 = 64 KiB (A16,W4)") reduces energy consumption down to an 18.4%—with respect to the reduction already achieved by platform 06. However,

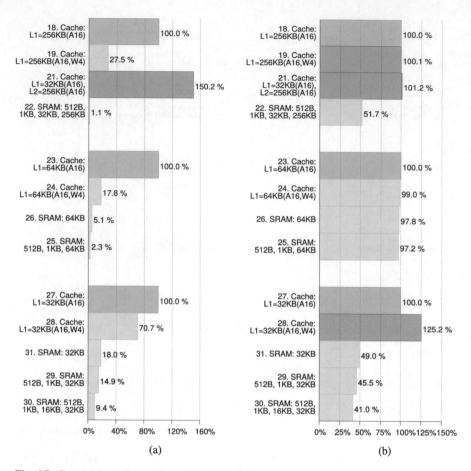

Fig. 6.7 Case study 1: Comparison of SRAM-based solutions with solutions based on caches of equivalent capacity for platforms with LPDDR2-SDRAM. Each group of results is normalized to the uppermost bar (in blue), which represents the performance for a cache of a given size. Red bars mark configurations that have worst performance than their reference. (**a**) Energy consumption. (**b**) Cycles memory subsystem

using instead an SRAM of the same size (platform "09. SRAM: 64 KiB"), the energy consumption is reduced down to 5.9% of the energy consumption with the 64 KB cache. Furthermore, this still represents a reduction in energy consumption of a 68.4% from platform 07 (the one with shorter cache lines). Interestingly, platform "08. SRAM: 512 B, 1 KiB, 64 KiB" shows that SRAMs can be seamlessly composed into hierarchies to achieve even greater improvements. Our methodology generates automatically solutions to exploit these complex memory hierarchies.

Similar trends are shown in Fig. 6.7a for the platforms with LPDDR2-SDRAM. One interesting observation is that the relative improvement of platform "28. Cache: L1 = 32 KiB (16,W4)" with respect to platform "27. Cache L1 = 32 KiB (A16)" is smaller

than in the case of the platforms with Mobile SDRAM. This effect is likely due to
the prefetch nature of LPDDR2, which eases the transfer of long cache lines while
eliminating the difference between single and double word transfers.

Interestingly, Figs. 6.6b and 6.7b show a different behavior for the number of
cycles than for the energy consumption. These differences owe to two factors. First,
latencies do not scale linearly with the size of the caches. For instance, we used
a latency of one clock cycle for all the memories smaller than 128 KB, be them
caches or SRAMs; also, we assumed that SRAMs and caches of the same size have
the same latency. And, second, latencies can be hidden in some cases. For example,
in the case of write latencies not immediately followed by a read. In that way, some
of the additional accesses incurred by the cache platforms do not directly affect
the performance of the system. However, energy consumption cannot be hidden:
If a superfluous memory access is performed, the system will always pay the due
energetic cost. In general, we can conclude that the platforms with SRAMs managed
with our data placement methodology have better performance.

6.4   Case Study 2: Network Routing

In this case study we will explore the application of the methodology to a network
dispatcher based on the deficit round robin (DRR) algorithm. The system is
implemented as a multithreaded application. Therefore, several instances of the
application DDTs are alive and accessed at the same time by different threads;
hence, the sequentiality of memory accesses is broken because accesses from
different threads interleave in a non-deterministic way. Indeed, if the system runs
on a multiprocessor platform—case that is not covered in this work—the accesses
will be interleaved every few words. For example, if two DDTs accessed with a
sequential pattern by different threads are placed on the same bank of a DRAM,
these access sequences, that would otherwise execute efficiently, will generate a big
number of different row activations.

6.4.1   Description of the Application

Embedded systems may execute threads from several applications concurrently;
the operating system has then to arbitrate between the threads that need to send
data through the network. A common choice in these cases is deficit round robin
(DRR) [9], which is a network fair scheduling algorithm that splits the available
bandwidth evenly among a number of destinations, The operating system keeps a
list of active destinations and assigns a quota to each of them, possibly taking into
account relative priorities. When a thread sends a packet, it is stored in the queue
corresponding to that network destination. Packets are extracted from the queues
in order and forwarded to the network adaptor, reducing the credit of the queue

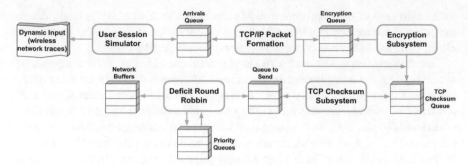

Fig. 6.8 Structure of the application used in the second case study. The boxes represent the different threads/kernels that communicate through asynchronous FIFO queues

proportionally to the size of the packet. DRR enables the implementation of quality-of-service (QoS) mechanisms that prioritize among applications and destinations, for instance, to guarantee a minimum bandwidth for certain connections while avoiding starvation in the rest.

The system implementation is organized in five modules that communicate through asynchronous FIFO queues, as illustrated in Fig. 6.8:

- Packet injection: To simulate traffic generated by client applications, this sub-system uses a collection of real wireless network traces to inject packets in the system.
- Packet formation: Using data from the network traces, a TCP/IP header is added to each packet that enters the system.
- Encryption: Packets that belong to encrypted sessions are passed through this subsystem to simulate the work of a block cypher.
- TCP checksum computation.
- Scheduling: DRR classifies the packets in priority queues and schedules them according to available bandwidth.

In order to reproduce the conditions of a real system, we use a set of network traces collected from the wireless access points in the campus of the Dartmouth University [6]. We identified traces from individual, yet anonymous, users and applications; some of them represent sessions lasting a few minutes while others represent sessions of up to 24 h. This allows reproducing part of the original system use cases, with the exception of accurate packet rate control.

6.4.2   Experimental Results

We can see in Table 6.4 the results obtained after applying our methodology, in comparison with the ones obtained with cache memories of similar sizes. In total, we have evaluated 32 different platform configurations with 14 network traces. The

Table 6.4 Case study 2: Performance of the solutions obtained with our placement methodology versus cache-based solutions

Platform	Energy	σ_{n-1}	Time	σ_{n-1}	Page misses	σ_{n-1}	DRAM accesses	Total accesses
(Mobile SDRAM)								
00. Only DRAM	100.0	0.0	100.0	0.0	100.0	0.0	100.0	100.0
01. Cache: L1 = 256 KiB (A16)	46.6	30.3	39.5	26.1	7.2	3.5	43.2	184.1
02. Cache: L1 = 256 KiB (A16,W4)	38.9	25.6	40.9	23.5	16.5	7.0	41.3	173.1
03. Cache: L1 = 512 KiB (A16)	42.8	37.1	34.4	29.6	4.3	4.8	33.5	164.9
04. Cache: L1 = 4 MB (A16)	107.0	66.5	54.7	36.2	3.1	4.3	28.6	155.5
05. Cache: L1 = 16 KiB (A16), L2 = 256 KiB (A16)	51.4	34.4	41.2	29.7	7.2	3.5	43.2	292.4
06. Cache: L1 = 32 KiB (A2), L2 = 256 KiB (A16)	48.6	33.4	41.3	29.7	7.2	3.6	43.2	293.5
07. SRAM: 512 B, 1 KiB, 32 KiB, 256 KiB	25.0	25.8	22.7	19.6	0.8	0.9	31.4	100.0
08. SRAM: 256 KiB	25.9	25.6	25.8	18.8	0.9	0.9	32.0	100.0
09. SRAM: 4MB	15.3	14.3	36.5	12.9	0.3	0.4	12.6	100.0
10. SRAM: 8 × 512 KiB	11.3	14.9	17.6	11.8	0.3	0.4	12.6	100.0
11. SRAM: 512 B, 1 KiB, 32 KiB, 256 KiB, 8 × 512 KiB	10.1	13.9	14.2	11.9	0.3	0.4	11.7	100.0
12. SRAM: LowerBound	<0.1	<0.1	5.7	1.7	0	0	0	100.0
13. Cache: LowerBound(D)	6.9	3.2	7.0	2.5	<0.1	<0.1	2.7	105.3

(continued)

Table 6.4 (continued)

Platform	Energy	σ_{n-1}	Time	σ_{n-1}	Page misses	σ_{n-1}	DRAM accesses	Total accesses
(LPDDR2-SDRAM)								
14. Only DRAM	100.0	0.0	100.0	0.0	100.0	0.0	100.0	100.0
15. Cache: L1 = 256 KiB (A16)	47.9	36.4	34.9	24.6	5.2	2.8	43.2	184.1
16. Cache: L1 = 256 KiB (A16,W4)	57.4	50.8	56.9	44.5	10.1	5.1	41.3	173.2
17. Cache: L1 = 512 KiB (A16)	50.8	44.3	32.0	26.4	2.5	3.0	33.5	165.0
18. Cache: L1 = 4 MB (A16)	187.6	139.2	64.0	42.5	1.4	2.0	28.6	155.5
19. Cache: L1 = 16 KiB (A16), L2 = 256 KiB (A16)	58.5	47.3	38.2	30.5	5.2	2.8	43.2	292.4
20. Cache: L1 = 32 KiB (A2), L2 = 256 KiB (A16)	53.0	44.0	38.2	30.5	5.2	2.8	43.2	293.6
21. SRAM: 512 B, 1 KiB, 32 KiB, 256 KiB	12.7	13.4	12.9	8.1	0.9	0.9	31.4	100.0
22. SRAM: 256 KiB	13.7	13.4	16.7	7.9	0.9	0.9	32.0	100.0
23. SRAM: 4MB	15.6	10.3	43.7	15.3	0.3	0.4	12.6	100.0
24. SRAM: 8 × 512 KiB	7.7	9.5	17.0	8.4	0.3	0.4	12.6	100.0
25. SRAM: 512 B, 1 KiB, 32 KiB, 256 KiB, 8 × 512 KiB	6.6	8.9	13.2	8.7	0.2	0.4	11.7	100.0
26. SRAM: LowerBound	0.1	<0.1	8.4	4.1	0	0	0	100.0
27. Cache: LowerBound(D)	10.9	4.5	9.1	4.2	<0.1	<0.1	2.7	105.3

Average normalized improvements with sample standard deviations (σ_{n-1}, sample size $N = 14$). All figures are percentages. The entries in the first half of the table, which correspond to platforms with a Mobile SDRAM, are normalized using as reference platform 00. The entries in the second part (platforms with an LPDDR2) are normalized taking as reference platform 14

results for every platform with every input are normalized against the results of the reference platform with the same input, and the normalized values are then averaged for each platform. As an example, platform 07 consumes a 25.0% of the energy consumed by its reference platform (00) across all the input traces.

The sample standard deviation in the table shows significant fluctuations from the average values for almost every platform because of the different nature, transmitted data length, and duration of the inputs. However, the solutions generated with the methodology improve always on the results obtained with caches. Interestingly, the standard deviation of our solutions is usually smaller than that of cache solutions, suggesting a more uniform system performance.

In this experiment, memory footprint varies approximately from 242 KB to 8.7 MB, with 13 out of the 14 cases over 512 KB. This situation attests that the solutions produced by our methodology have good performance also when the application footprint exceeds the size of the available on-chip SRAMs. The reason for this good performance is that the data placement puts in the DRAM only instances of the least accessed DDTs and, in this case, those instances (the body of the network packets) are accessed mostly sequentially. Temporal locality is low because each instance is accessed just twice, first to calculate the CRC and then to forward it to the buffers of the network adaptor. As our solutions can split the pools over different memory resources, some of the packet-body instances are still placed on the on-chip SRAMs, but no instances of other more accessed DDTs are evicted.[1] If the application has a memory allocation pattern that alternates peaks with periods of lower consumption then, during a potentially significant fraction of the execution time, all the instances of all the DDTs may reside in the on-chip memories without conflicts (that is, if the footprint of the packet bodies is small enough to fit in the part of their pool mapped on the closer memories). As the number of packet bodies increases, their memory pool overflows and some of them are allocated in the DRAM. However, the crucial difference with respect to cache platforms is that accessing those objects does not evict more accessed data from other pools. In this situation—accessing big data objects with little reuse—cache memories cannot amortize the cost of data movements and even risk evicting more useful data. An alternative in platforms based on caches could be placing these DDTs in a non-cacheable area, effectively deactivating caching for all the packet bodies.

In a related consideration, we can observe an increase in energy consumption for platforms with bigger cache memories. This is due to the fact that bigger caches consume more energy per access, and the smaller ones are already capable of the most significant reduction of accesses to the DRAM. This effect is less clear in the number of cycles required for execution because the cache latencies used in

---

[1]It is possible to split the pool of packet bodies in two areas, one cacheable and the other non-cacheable, allocating space from the first one as long as possible. However, how big should the cacheable pool be? The answer to this question would probably require an analysis similar to the one proposed in this text!

the experiments do not increase linearly with their size. Indeed, cache hierarchies perform quite badly in this experiment. Let us consider, for example, the case of platforms 01 and 05. Despite having more capacity, the second platform has considerably higher energy consumption. This is due to the continuous transfers of data with low locality between the small L1 and the L2. As a result, the total number of memory accesses, that is, the addition of accesses to all the memory modules, including transfers between levels in the cache hierarchy, is much higher in the second platform. Compare these results with the ones obtained for platform 08, which has an SRAM of the same size than the cache of platform 01.

Another interesting observation is that the values of energy consumption shown in Table 6.4 are compared independently for the platforms with Mobile SDRAM and LPDDR2. However, a direct comparison between them exposes a net reduction of a 39.8% on an average for all platforms ($s = 26.7\%$) when using an LPDDR2 instead of the older technology (specifically, 71.0% with $s = 8.3\%$ when comparing platforms 21 and 7, excluding the experiments that fit entirely in the on-chip memories). This highlights the important effort invested by the industry in reducing the energy consumption of the memories designed for embedded systems.

To conclude our experiment, in Table 6.5 we can observe non-aggregated data for several input cases to compensate for the big standard deviation in the aggregated results of Table 6.4. With this information, we can more easily compare the performance of our solutions for several SRAM-based platforms with standard configurations that include caches of equivalent sizes.

6.5   Case Study 3: Synthetic Benchmark—Dictionary

The goal of this last example is to show that DM-intensive applications can limit the effectiveness of cache memories because of their low spatial and temporal localities. This benchmark uses a trie to create an ordered dictionary of English words, and then performs multiple look-up operations. The trie DDT [4] belongs to the category of ordered trees and is useful to store any type of information that can be organized using prefixes, especially if it presents a high degree of redundancy, such as words of a dictionary, compression tables, and DNA sequences. Here, each node has a list of children indexed by letters.

This experiment models a case that is particularly hostile to cache memories because each traversal accesses a single word on each level, the pointer to the next child, but the cache has to move whole lines after every conflict. This is a well-known side effect of the use of dynamically linked data structures on cache architectures, including desktop and server computers, and is thus an area of intense research.

Figures 6.9 and 6.10 show the improvements attained with our methodology in comparison with cache memories. As before, each of the figures shows three different comparisons, where the results of each group are normalized to the value of the first bar in the group (in blue). In both figures, the first group of bars compares the

Table 6.5 Case study 2: Detailed comparison between SRAM and cache-based solutions

Platform	(Mobile SDRAM)				(LPDDR2-SDRAM)			
	Energy	Time	DRAM accesses	Page misses	Energy	Time	DRAM accesses	Page misses
Input 1								
Cache: L1 = 256 KiB (A16)	100.0	100.0	100.0	100.0	100.0	100.0	100.0	100.0
Cache: L1 = 64 KiB (A4)	98.9	92.9	123.8	130.2	79.6	80.4	123.9	141.1
SRAM: 64 KiB	52.4	56.7	79.5	30.6	45.8	61.1	79.6	43.3
SRAM: 256 KiB	24.3	53.2	32.5	2.5	16.9	61.4	32.6	3.1
SRAM: 512 B, 1 KiB, 32 KiB, 256 KiB	19.7	35.5	28.7	1.9	11.4	37.5	28.7	2.2
Input 2								
Cache: L1 = 256 KiB (A16)	100.0	100.0	100.0	100.0	100.0	100.0	100.0	100.0
Cache: L1 = 64 KiB (A4)	94.1	93.5	104.8	131.1	80.7	83.1	104.8	134.7
SRAM: 64 KiB	72.3	68.8	95.9	40.1	38.7	41.8	95.9	71.9
SRAM: 256 KiB	68.3	68.2	90.7	14.5	33.2	41.4	90.7	20.1
SRAM: 512 B, 1 KiB, 32 KiB, 256 KiB	67.6	65.0	90.2	14.4	32.6	36.1	90.2	20.0

(continued)

Table 6.5 (continued)

Platform	(Mobile SDRAM)				(LPDDR2-SDRAM)			
	Energy	Time	DRAM accesses	Page misses	Energy	Time	DRAM accesses	Page misses
Input 3								
Cache: L1 = 256 KiB (A16)	100.0	100.0	100.0	100.0	100.0	100.0	100.0	100.0
Cache: L1 = 64 KiB (A4)	86.5	83.0	103.9	102.9	71.7	72.9	103.9	102.3
SRAM: 64 KiB	46.8	50.9	69.5	20.2	42.6	56.7	69.5	29.7
SRAM: 256 KiB	22.2	49.1	27.8	5.2	17.7	59.3	27.8	7.7
SRAM: 512 B, 1 KiB, 32 KiB, 256 KiB	17.8	32.8	24.2	4.7	12.6	37.1	24.2	6.9
Input 4								
Cache: L1 = 256 KiB (A16)	100.0	100.0	100.0	100.0	100.0	100.0	100.0	100.0
Cache: L1 = 64 KiB (A4)	92.4	91.9	102.8	122.4	79.4	81.5	102.8	119.9
SRAM: 64 KiB	73.2	69.5	96.8	38.6	37.8	40.3	96.8	78.5
SRAM: 256 KiB	70.3	69.6	92.8	20.3	34.2	41.3	92.8	34.6
SRAM: 512 B, 1 KiB, 32 KiB, 256 KiB	69.7	66.8	92.5	20.2	33.1	36.1	92.5	34.4

The execution cost of the solutions obtained with our methodology is normalized for each input against the cost of a solution with a cache memory of equivalent size. These results correspond to 4 of the 14 inputs used in the experiments, not aggregated. All numbers are percentages

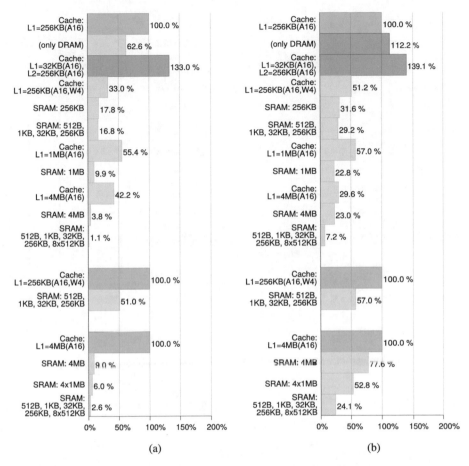

Fig. 6.9 Case study 3: Comparison of SRAM-based solutions with solutions based on caches of equivalent capacity for platforms with Mobile SDRAM. Each group of results is normalized to the uppermost bar (in blue), which represents the performance for a cache of a given size. Red bars mark configurations that have worst performance than their reference. (**a**) Energy consumption. (**b**) Cycles memory subsystem

performance of caches and SRAMs of increasing sizes to that of a 256 KB cache. A direct comparison between platforms "SRAM: 512 B, 1 KB, 32 KB, 256 KB" and "Cache: L1 = 256 KB (A16)" shows that our solution achieves a relative improvement of 83.2% for the LPSDRAM case and 77.8% for the LPDDR2 case in energy consumption, 70.8% and 60%, respectively, when considering the cycles spent accessing the memories. As discussed for the previous case study, cache hierarchies (platform "Cache: L1 = 32 KB (A16), L2 = 256 KB (A16)") perform poorly with this type of applications because the small size of the caches in the closer levels forces many evictions of whole lines, and the transfers between cache levels accumulate to the accesses to the DRAM themselves.

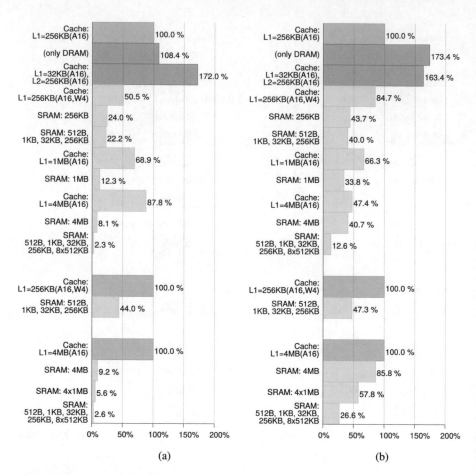

Fig. 6.10 Case study 3: Comparison of SRAM-based solutions with solutions based on caches of equivalent capacity for platforms with LPDDR2-SDRAM. Each group of results is normalized to the uppermost bar (in blue), which represents the performance for a cache of a given size. Red bars mark configurations that have worst performance than their reference. (**a**) Energy consumption. (**b**) Cycles memory subsystem

We can observe a very interesting result reducing the size of the cache lines. Let us consider as baseline the configuration offered by many processors such as the ARM Cortex-A15, which has 64-byte cache lines. Although the trend for larger line sizes may be beneficial for applications that process streams of data (because it increases the extent of prefetching), for applications that rely heavily on the use of DDTs the effect is quite different: Every line fill and write-back is more costly because more words are moved between memories even if the application accesses just one or two of them, and the longer lines reduce the number of different cache lines (i.e., different memory positions) that can be stored with the same cache size. In our results we can see that using a line size of just 16 B (4 words) can improve, for this type of applications, energy consumption by 67% with LPSDRAM and 49.5% with LPDDR2, and the cycles spent in the memory subsystem by 48.8% and 15.3%, respectively (platform "Cache: L1 = 256 KB (A16,W4)" versus "Cache: L1 = 256 KB (A16)"). Nevertheless, our solutions with explicit data placement on addressable memories are even better suited for this type of applications, still improving energy consumption over the 16-byte-lines cache memory by 49% and 56%, and the cycles used by 43% and 52.7%, respectively (platform "SRAM: 512 B, 1 KB, 32 KB, 256 KB" versus "Cache: L1 = 256 KB (A16, W4)," second block of bars in the figures).

Finally, the third group of bars in the figures shows that the performance of our solutions also scales better with the size of the memories. Moreover, this last group of bars shows how multiple on-chip SRAMs of a smaller size can be combined to further improve performance, in contrast with the difficulties encountered while trying to harness cache hierarchies.

6.6 Additional Discussion

6.6.1 Suitability and Current Limitations

The data placement methodology that we have explored throughout this text is in its current form best suited for applications that use DDTs in phases or traverse data structures accessing only a small amount of data at each node, and whose DDTs have very different FPBs. These cases may hinder the performance of hardware caches as they have to move more data around the memory hierarchy than is really needed, with the additional issue of possibly evicting very accessed objects with seldom accessed ones. As an example, the traversal of a structure could force a cache to move complete nodes back and forth, with an increasing waste of energy as the size of the cache lines increases. The situation can be improved splitting of data structures to pack the node pointers tightly and access only the data of the nodes that are actually needed during the traversal. This effect is particularly beneficial with our methodology because the pool containing the pointers is guaranteed to be always in the correct memory module, independently of which other data accesses are performed by the application.

On the contrary, our methodology may not be adequate for applications that keep instances of many DDTs alive simultaneously and alternate between phases that access each of them, particularly if each phase creates periods of high access locality. The reason is that the objects do not free their space and thus, grouping cannot reuse it. In comparison, cache memories are specifically designed for this situation: They use data movements to recycle storage and keep the most currently accessed data closer to the processor. For these situations, cache memories have proven themselves useful during decades. One exception to this consideration arises when the application has low spatial locality because our methodology avoids moving data words that will not be reused. Even if accessing the objects of the group in the worst resource incurs a high cost, saving fruitless data movements may produce important energy savings.

Additionally, our methodology assumes that most instances of a given DDT have a similar FPB. Therefore, it may not be adequate for applications whose DDTs have instances with very different FPB. Moreover, it leaves the least accessed DDTs in the most distant memory modules; this is appropriate for data streams that are processed sequentially and without reuse, or for small data elements that are seldom accessed. However, for other access patterns to big arrays, it may be convenient to supplement our approach with software techniques such as array tiling or blocking on a small dedicated SRAM.

As a final important consideration, the performance of the solutions generated with our methodology degenerates in the worst case to the performance of a platform with only DRAM memories (or the most inefficient technology used in the memory subsystem, if other). A simple justification is that accesses happen either to an SRAM or to the DRAM, where any access to an SRAM improves performance. As the total number of accesses does never increase—in stark contrast with cache hierarchies—the total cost is bounded by the cost of executing all the accesses on the DRAM (small variations might nonetheless happen because of specific data layouts that produce slightly different numbers of DRAM row-misses, even if the total number of accesses to the DRAM is reduced).

6.6.2   Use Cases for the Methodology

Our methodology can be used in two different ways, depending on whether the hardware platform is fixed or not at the time of analysis:

6.6.2.1   Application Optimization for a Fixed Platform

When the hardware platform is fixed, the methodology can be used to produce the placement of the dynamic data objects of the application into the available memory resources and improve energy consumption and performance. It may also help to increase the extent of the periods during which the external DRAMs are on low-power modes.

6.6.2.2   Hardware Platform Exploration and Evaluation

The system designer can use the simulator that we have presented to explore the performance of the application on different memory hierarchies and choose the most suitable one from the available options. Alternatively, if the design is going to be implemented on an ASIC (or to a lesser extent on an FPGA), the designer may have complete control to adapt the platform exactly to the characteristics of the application, maybe including multiple small memories.

An interesting possibility to explore is integrating in the design as many memory modules as groups defined by the tool, with the size of each one close to the size of the corresponding group. The size and FPB of the groups may help to decide the size of the memories. For example, the designer may instantiate a small SRAM to contain6 a small group with a high FPB and enjoy the lower energy per access required by very small memories—indeed, recent advances in logic synthesis may even enable the use of standard cell memories (SCM) for such tiny memories [10]. Bigger groups with similar FPBs may be included in a single SRAM, or the designer may instantiate separate ones, depending on the trade-off between control logic overhead and the energy savings—and possibly higher bandwidth—introduced by having smaller memories, and, perhaps, the possibility of shutting down some of them during specific phases of the application. Figure 6.11 explores those possibilities briefly. The first group of bars shows the case of 4 DDTs, each with a footprint of 1 KB, mapped either together on one SRAM of 4 KB or separately on four SRAMs of 1 KB each. The second group of bars shows the case of 4 DDTs of 4 KB each, mapped either on one 16 KB SRAM or on four 4 KB ones. Finally, the third group shows the case of 2 DDTs of 32 KB each, mapped on either one 64 KB SRAM or on two 32 KB ones. Assuming that each DDT receives an equal number of 5×10^5 accesses along the execution time of the application, we can derive the expected energy consumption (Fig. 6.11a) and the area overhead of having multiple memories each with its own control logic (Fig. 6.11b).

Increasing the number of elements in the memory subsystem may introduce some overheads because of the inclusion of multiple address decoding elements and less regular geometries (although probably partially offset by the lower cost of the memory decoders), but it does not introduce any of the complexity issues that arise with cache hierarchies. This is especially true for "miss-hit" chains in cache hierarchies, which are completely absent in the solutions produced with our methodology.

Other design aspects that can be explored are the advantages of dividing DRAMs into more banks, or using more DRAM modules instead of a bigger single one to increase the number of banks that can be active at the same time. Or the possibility of introducing additional SRAM capacity in the design with the advantage of reducing DRAM size or even completely removing it. For example, if an application uses exactly 257 MB of memory, the designer may evaluate including a 1 MB SRAM in the ASIC instead of using a bigger DRAM. While such a decision may be out of question with caches (the system will still require the 257 MB of DRAM), our methodology enables it and even takes care of the data placement issues transparently.

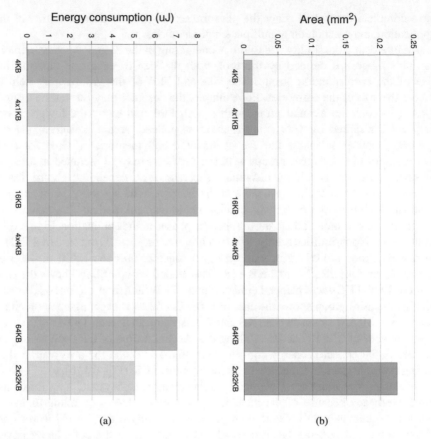

Fig. 6.11 With our simulator, we can easily explore the repercussions of splitting the SRAMs of the system into several smaller memories, both on energy consumption reduction (**a**) and area overhead (**b**)

6.6.3 Simulation Versus Actual Execution

Although we have put as much care and attention as possible into the design of our memory hierarchy simulator—and not less on trying to give enough information to the reader in order to understand it—multiple factors may affect the accuracy of the results that we have obtained with it. Thus, it is important to keep in mind that simulation is just a convenient way to assess the properties of the placement solutions generated with the methodology, or to evaluate multiple platform variations quickly.

Consequently, additional experimentation on real hardware platforms is required to validate the improvements that the experiments presented in this chapter promise. Nevertheless, we believe that the significant improvement margins obtained for most cases support the plausibility of the techniques presented throughout this work.

6.6.4   Reducing Static to Dynamic Data Placement—or Vice Versa

The techniques for dynamic data placement presented in this text can be applied also to generate a placement for static data. The implementation would be technically easy: Each global variable or object declaration is simply replaced by a pointer declaration. Then, a function call is included at the start of the application to allocate all these objects via calls to `malloc()` or the `new` operator. With this simple change, the properties of each variable can be analyzed with our methodology as a DDT with a single instance (i.e., a "singleton") and considered during placement. Thus, static global objects will be allocated at run-time in SRAM or DRAM resources according to their access characteristics together with the other objects of the application. Moreover, further improvements in our methodology may also help to separate global data objects into different DRAM banks without explicit attention. The designer may exclude explicitly these pseudo-DDTs from grouping to ensure that they are not mixed in pools with other objects, but this is probably unnecessary because they will be allocated at the beginning of the execution and never destroyed.

An important advantage that can be obtained by managing global static data objects with our methodology is adaptability to platform changes. For example, if the application is executed in a platform with bigger SRAMs, static data objects will also benefit from them according to their characteristics. Similarly, in platforms with smaller resources the complete data placement solution will be automatically reevaluated for all the data objects. More importantly, the system will be able to run even if the SRAM in which a global data object would have been mapped fails.

Stacks may also be considered as dynamic arrays and placed accordingly (depending on the concrete processor architecture requirements), but other existing approaches that include specific hardware support [5] seem more adequate. Special requirements such as stack management during interruptions may prevent the applicability of our techniques to these data objects.

In the other direction, it is also possible to use existing techniques for placement of static data objects instead of our mapping phase. As explained at the beginning of this text, the problem lies not so much on placing the pools themselves on memory resources, but on classifying the dynamic data objects of the application into different pools: Creating a big pool for all the objects would blindly mix very and seldom accessed objects, hence making differentiation impossible. With hardware caches, this solution would correspond to the model traditionally used in desktop computers. Software caching techniques would not be very efficient due to the impossibility of identifying individual objects in the pool address space and the lack of locality. However, after the DDTs are classified with a mechanism such as our grouping step, it would be possible to place each complete pool into memory resources as if it were a static array. Software caching techniques might also be applied to specific pools that contain dynamic data objects with particular characteristics.

In a sense, both problems are complementary with plausible reductions in both directions, although these reductions may introduce some overheads.

6.6.5   Order Between Mapping and Pool Formation

In the current way in which our methodology is formulated, pool formation, that is, the design of the dynamic memory manager for each pool, is performed at design time before the mapping step. The reason is that this step is usually complex, frequently requiring some exploration of the design space. One of our main goals was that the mapping step can be moved to run-time so that the application can be adapted to the concrete resources available in the platform at the time of execution, maybe with variations between successive executions. Thus, any parts of the methodology executed at run-time must be very efficient. This decision may be reconsidered if efficient techniques for the design of dynamic memory managers are found.

Traditionally, DMMs have been designed taking into account the number and size of the allocated blocks and the allocation and deallocation patterns of the application. However, it may be possible to improve the performance of the DMM itself by taking into consideration the characteristics of the memories where the pools are placed. For example, a DMM for a pool located in an SRAM may employ aggressive coalescing and splitting mechanisms to reduce fragmentation at the expense of some more—relatively cheap—memory accesses. In contrast, a DMM for a pool located in a big DRAM may relax anti-fragmentation measures to restrain the number of costly random memory accesses.

To exploit this information, pool formation must be executed after the pools are placed on memory resources. Therefore, either the mapping step is always executed at design time or pool formation is delayed until run-time. The alternative, if no efficient techniques for DMM design are available, may be to generate at design time multiple DMM descriptions, one for each type of memory technology that deserves special attention. Then, the mapping step could be executed at design or run-time as convenient. Assuming that this does not constitute a big burden for deployment, it may be an interesting option for future research. Nevertheless, the benefits of this order change still need to be assessed.

References

1. ARM: Cortex-A15 Technical Reference Manual Rev. r2p0. ARM (2011)
2. CACTI 5.3 (2008). http://quid.hpl.hp.com:9081/cacti/
3. Catthoor, F., Raghavan, P., Lambrechts, A., Jayapala, M., Kritikakou, A., Absar, J.: Ultra-Low Energy Domain-Specific Instruction-Set Processors, 1st edn. Springer Netherlands (2010). https://doi.org/10.1007/978-90-481-9528-2

4. Fredkin, E.: Trie memory. Commun. ACM **3**(9), 490–499 (1960). https://doi.org/10.1145/367390.367400

5. González-Alberquilla, R., Castro, F., Piñuel, L., Tirado, F.: Stack filter: reducing L1 data cache power consumption. J. Syst. Archit. **56**(12), 685–695 (2010). https://doi.org/10.1016/j.sysarc.2010.10.002

6. Henderson, T., Kotz, D., Abyzov, I.: The changing usage of a mature campus-wide wireless network. In: Proceedings of the Annual International Conference on Mobile Computing and Networking (MobiCom), pp. 187–201. ACM Press, Philadelphia (2004). https://doi.org/10.1145/1023720.1023739

7. Hennessy, J.L., Patterson, D.A.: Computer Architecture: A Quantitative Approach, 5th edn. Morgan Kaufmann, San Francisco (2011)

8. Ingelrest, F., Barrenetxea, G., Schaefer, G., Vetterli, M., Couach, O., Parlange, M.: SensorScope: application-specific sensor network for environmental monitoring. ACM Trans. Sensor Netw. **6**(2), 1–32 (2010). https://doi.org/10.1145/1689239.1689247

9. Shreedhar, M., Varghese, G.: Efficient fair queuing using deficit round-robin. IEEE/ACM Trans. Netw. **4**(3), 375–385 (1996). https://doi.org/10.1109/90.502236

10. Teman, A., Rossi, D., Meinerzhagen, P., Benini, L., Burg, A.: Power, area, and performance optimization of standard cell memory arrays through controlled placement. ACM Trans. Design Autom. Electron. Syst. **21**(4), 59:1–59:25 (2016). https://doi.org/10.1145/2890498

Chapter 7
Closing Remarks and Open Questions

7.1 Summary of Main Ideas

Throughout all this text we have motivated the need for specific techniques for the placement of dynamic data objects in systems with heterogeneous memory organizations and the important benefits that they can bring. In particular, we have advocated a static and exclusive placement, implemented through the dynamic memory manager, that avoids data movements for suitable dynamic data types. The following list summarizes the main conclusions of this text:

- The use of dynamic memory and dynamic data structures typically lowers the access locality of applications, hindering the performance of caching techniques. Specific methods for placement of dynamic data objects that limit or avoid data movements can improve the performance and reduce the energy consumption of these applications, especially when they are executed on systems with heterogeneous memory organizations.
- We have proposed a systematic placement methodology for dynamic data objects that avoids movements across elements in the memory subsystem and exploits the characteristics of individual memory modules to improve accesses to dynamic data objects—or to avoid that dynamic data objects hinder the normal work of caching techniques for other data objects. An important consideration is balancing between exclusive assignment of resources and resource exploitation to avoid saving data movements at the price of wasted resources.
- The problem of dynamic data placement is complex. To tackle with that complexity, we have divided it into several phases: Preliminary characterization and analysis, grouping of DDTs with similar characteristics, and mapping of those groups into actual memory resources. The first two steps are complex and should be performed during design time, but the last one may be delayed until run-time to improve system adaptability to resource degradation or different platform configurations.

© Springer Nature Switzerland AG 2020 175
M. Peón Quirós et al., *Heterogeneous Memory Organizations*
in Embedded Systems, https://doi.org/10.1007/978-3-030-37432-7_7

- The methodology requires an extensive characterization of the application, so a preliminary profiling phase is required. The same instrumentation can also supply the information needed by the dynamic memory manager; hence, the effort is shared for both purposes.
- The grouping step is a trade-off that provisions dedicated space for the instances of the most accessed DDTs of the application while limiting resource underutilization. It combines DDTs whose instances present equivalent access characteristics, or that have complementary footprint demands so that when the space is needed for highly accessed instances most of the less accessed ones have already been destroyed and the space is again available.
- The second phase, mapping of the generated groups of DDTs into memory resources, is platform-dependent. It is also computationally simpler because it is a particularization of the integer knapsack problem. Thus, this step may be delayed until run-time to automatically configure the system for various platform configurations or to achieve graceful performance degradation as it ages and some components start to fail.
- The dynamic memory manager (DMM) can be the means to implement the generated placement solutions. To carry out this new responsibility, the DMM requires extra information: The type of the data object involved in each allocation operation.
- A memory simulator is required to explore the performance and energy consumption of the application running on different heterogeneous memory organizations with specific placement solutions. This exploration can be used to tune the application to an existing platform, or to explore the performance of different platform architectures subject to representative applications and data placements.
- The results of the experiments show that the placement of dynamic data objects produced by the methodology clearly improves the performance and energy consumption in comparison with the utilization of traditional cache memories, without adding expensive HW or SW requirements. The cause of the improvements is not only that SRAMs are usually more efficient than caches built with the same technology, but also a reduced number of memory accesses.

To conclude, we would like to remark that the importance of this work is not so much on the concrete algorithms presented throughout this text as on the necessity of considering data placement at all the abstraction levels, from the nodes of a linked list in a simple application to complex data repositories in dedicated rack-level memory resources, and on the overall methodology—particularly its organization into steps and their relative ordering.

The research presented in this text was part of a global project that spawned several institutions and generations of researchers, led by Professors Francky Catthoor (IMEC, Belgium), Dimitrios Soudris (DUTH and currently NTUA, Greece), and José Manuel Mendías José Manuel Mendías Cuadros (UCM, Madrid).

7.2 Open Questions

In the following paragraphs we discuss briefly some topics worth of future research. First, we enumerate open avenues for research on each of the steps of the methodology for placement of dynamic data objects. Then, we consider the applicability of our proposals to other computing environments besides embedded systems.

7.2.1 Methodology and Algorithms

The methodology presented in Chaps. 4 and 6 is based on the heuristic of splitting the placement problem in two steps, grouping and mapping. However, it is conceivable that both steps are performed at the same time, even with a perfect solver (possibly based on estimators for partial solutions), to evaluate the computational complexity of that approach and its possible improvements.

7.2.1.1 Profiling

The profiling methods that we have presented in this text have two main drawbacks. The first one is the necessity for manual instrumentation of the source code. Even with the reduced overhead of the exception-based mechanism, the designer still has to modify the declaration of every dynamic class. Then, these methods are quite slow, hindering the analysis of time-sensitive applications.

One approach that could be interesting is the modification of the compiler so that it inserts automatically the instrumentation needed for memory access tracking and provides type information to the DMM API functions. Memory access instrumentation would be used only during profiling, but type information would be inserted also in the final code. As an example, LLVM [16] is a compiler infrastructure specifically built to ease its modification and currently used in multiple commercial and research products.[1] For example, type information is known to the compiler during a `new` operation; for `free` operations, it might be possible to obtain it using run-time type information (RTTI).

Other possibility would be to completely avoid profiling and extract equivalent information using other analysis techniques. For example, the number of accesses in each code branch could be determined using static analysis techniques; the frequency of execution of each branch could be then calculated with a lightweight profiling. Such methods would be especially important to apply the placement methodology to other environments with more complex applications as presented later in Sect. 7.3.

---

[1]Chris Lattner and Vikram Adve received the ACM Software System Award in 2012 for their work on LLVM http://awards.acm.org/award_winners/lattner_5074762.cfm.

Finally, the examples used in this text perform an initial profiling on a processor architecture that may not be the one used for execution. This is a very useful approach that can be used when the final platform is not yet available, but additional experiments should be conducted to analyze the deviations introduced by this approach. The most significant difference may be the availability of different numbers of general-purpose processor registers. With a reduced number of them, the compiler has to generate more accesses to memory. Although dynamic data objects are typically accessed through pointers—it is the pointers themselves which may be hold in registers—and traversals of data structures offer little opportunity for reuse, further experiments should be conducted to verify the impact of this approach. Particularly, because profiling on a workstation may be crucial in cases where the final platform does not have enough resources (storage, performance) to profile the application reasonably, or simply because a workstation may be much faster, or a cluster of them may be used to profile under a myriad of different conditions. An efficient ISA-level simulator may be an adequate option in those cases.

7.2.1.2 Grouping

Grouping is a mechanism introduced to improve resource exploitation while still being able to assign resources in exclusivity to the instances of some dynamic data types. Furthermore, the algorithm presented in Chap. 4 is based on a set of heuristics that may (and should) be improved in the future. The following paragraphs discuss several improvements worth of further examination.

The grouping algorithm can be extended to take into account the access pattern of the instances of each DDT: Some access patterns may be more suitable for specific types of memories. For instance, those DDTs with prominently sequential access patterns may be assigned to a DRAM, even if it is less efficient, whereas SRAM is reserved for other DDTs with more random access patterns that could hinder the row-oriented organization of DRAMs. In this regard, we conducted some preliminary experiments using the "selfishness" metric proposed by Marchal et al. [19, 20], which gives a measure of how sequential the accesses to a data structure are. With it, DDTs with a high selfishness can only be joined to groups that contain DDTs with a similarly high one. During mapping, pools with a high selfishness are mapped preferably on DRAM memories. The initial results were promising.

Regarding access pattern identification, we also conducted some preliminary tests in which DDTs whose objects are commonly accessed in tandem (e.g., dynamic vectors used in reduction operations) are marked as "incompatible," so that the grouping algorithm tries to avoid combining them and the mapping process avoids placing them in the same memory resources. For this to work, we augmented the attributes of DDTs, groups, and pools to include a list of incompatible peers (in the sense of preference for not being placed together, not of a strict incompatibility). These attributes can be used to avoid placing them in the same bank of a DRAM, thus reducing the number of row misses.

The possibility of placing two DDTs in different memories so that they can be accessed in parallel is a more complex topic that may require multiple data ports in the processor or at least interleaving of operations with long access times (DRAMs may fall in this category for random accesses, but they can transfer data continuously while in burst mode). An interesting study, limited to static data objects, is presented by Soto et al. [23]. This may constitute an interesting topic for future research in the context of dynamic data placement.

Perhaps, the most important aspect that should be investigated is the possibility of differentiating among instances of a DDT that present very different FPBs. Through this text, we have assumed that the instances of a DDT have all a similar number of accesses, but in some applications the situation may be different. An efficient answer would involve identification and specific placement of individual instances, or even migration of instances—maybe with mechanisms similar to those used by generational garbage collectors—once their specific characteristics are determined.

Other interesting experiment might be the modification of the grouping and mapping algorithms to consider the combination of a cache memory plus a DRAM as a single entity, something such as a "cached-DRAM." The idea is that the properties of the combination would be different than the properties of the DRAM. Instead of the approach presented in this text (leaving cached-DRAM areas for data objects that are not managed by the methodology) they would be seen as memories with special characteristics and automatically assigned by the methodology to the appropriate data objects, maybe opening the path to tackle as well the placement of more caching-amenable static data objects. The major obstacle for this approach would be that a simple statistic of typical hits and misses would not be enough: The methodology algorithms would need to know exactly how the cache reacts to the concrete (groups of) DDTs placed on its related memory and the possible interferences with other DDTs placed in the same DRAM cached area. Even more, multiple pools placed in the same cached-DRAM would interfere with each other. Therefore, this experiment might require also new algorithms to perform a simulation-guided full exploration of the design space.

7.2.1.3 Mapping

If, as suggested in Sect. 7.2.1.2, a mechanism to identify groups/pools that should be placed apart is added, then the mapping algorithm has to be modified to contemplate such restrictions, particularly to respect the placement of pools in different DRAM banks. Once the restrictions are generated during the grouping step, the modifications to the mapping algorithm might seem straightforward. However, some new issues arise for consideration.

For example, what should be done when a memory resource has still some free space, but the pools in it are incompatible with the new one being considered? One option is to skip that free space and map the pool in the next available memory resource. The next pool would use the space that was left in the first resource and the remaining space in the second one. Or would it be better to map the next pools until the old resource is fully used and then map the pool that was kept on hold? That option could produce a dangerous effect of "priority inversion" if the new pool considered is also incompatible with the one that was being kept on hold.

A third option would be modifying the mapping algorithm to pick the next pool that is compatible with the pools already mapped in the current memory module and map as much of it as possible. When the resource is exhausted, the pool would be inserted back in the list of pools, with its size adjusted, so that the algorithm would reconsider again all the available pools for the new resource. That option would observe pool priorities, but possibly producing many splits over quite distinct resources for some pools.

All of those options would be very easy to implement in $\mathcal{D}yn\mathcal{A}s\mathcal{T}$, but their effects should be carefully evaluated through additional experiments.

On a different topic, the most promising research path seems to be the execution of the mapping step at run-time. This approach promises very interesting possibilities for platform independence and system reliability via adaptation to resource degradation, making it a desirable research direction in the realm of embedded systems.

Although not strictly affecting the mapping phase, in Sect. 6.6.5 we analyzed the possibility of performing DMM design (pool formation) after the mapping step so that the actual properties of the memories assigned to each pool are known. The mapping step is agnostic with respect to the design of the DMMs in each pool; hence, that path could be pursued freely. However, one consideration is due: If executing the mapping step at run-time is deemed as interesting, then the pool formation step must also happen at run-time—strictly speaking, both would be executed at the time of loading the application in the same way than dynamic-library linking. We elaborate on this option below.

7.2.1.4 Integration in a Complete Dynamic Memory Management Framework

The design of dynamic memory managers has received a lot of attention during decades and thus, it is in a quite mature state. However, it would be interesting to see some meta-research on fast and low-resource techniques for the design of DMMs. The reason is that, traditionally, DMMs are built and evaluated during the design phase of the system. Consequently, even if reducing the time required to design a system is fundamental to reduce the time-to-market of embedded systems, time scales are completely different to what would be needed to perform DMM design at run-time after the mapping stage. For example, some prior work involves searches in the design space, whereas a run-time DMM design step would have to run in subsecond times. Run-time DMM design would motivate the development of quick heuristics for approximating a good DMM design given some predefined parameters.

In the absence of such quick design mechanisms, other options would still be possible. One would be executing pool formation normally at design time, but producing one different design for each possible type of memory resource in which the pool may be placed. Given that the number of different memory types is relatively small, the overhead should not be prohibitive, particularly if DMM designs are shipped in the form of templates for assemblage by a factory object. At run-time, the factory would choose the recipe that best matches the memory resource actually assigned to the pool—the normal approach explained in this text corresponds with the case of having only one recipe for each pool. Of course, before enrolling in such research, a previous step would be to effectively determine if the advantages obtained designing the DMMs considering the characteristics of the actual memory resources are indeed significant, and under what circumstances.

7.2.1.5 Deployment

In Sect. 4.11 we proposed the use of a library of modules to compose dynamic memory managers at run-time. However, an important consideration is that virtual calls to functions linked through a strategy design pattern imply a double indirection. Care has to be taken to determine the impact of this factor, especially for embedded processors with simple (if at all) branch predictors. If the overhead were too high, a more involved "patching" mechanism, similar to the work of the system loader for dynamic linking, could be explored on the basis that the DMM library would not be a completely arbitrary piece of code (i.e., the extent of the changes would be limited and the DMM objects would be created by the system itself, not freely by the application).

7.2.1.6 Simulation

Accuracy Simulation is a powerful tool to quickly analyze the characteristics of a system. For example, it allows the designer to explore different platform configurations fast and in a more flexible way than real execution, particularly if the physical platform is not yet available. However, its results may not be completely equal to those obtained during execution on a real system. Therefore, for each concrete platform, additional work should be conducted to assess the accuracy of the results obtained with the simulator compared with those obtained in the real platform.

Architectural Simulation We propose the use of a simulator based on memory traces. Although this allows analyzing the performance of the memory subsystem, some subtleties such as the timing between memory accesses are lost with this approach. An interesting future extension could be integrating the energy consumption and performance evaluation capabilities of the simulator with a full architectural simulation/emulation platform. In particular, integration with Gem5 [6] seems plausible and particularly promising: Gem5 can simulate several processor architectures and run complete systems with their operating system and applications via two working modes, System-call Emulation (SE) and Full-System (FS). Integrating a memory simulator such as ours with Gem5 would enable the analysis of the effect of multiple applications executed simultaneously or the impact of different scheduling policies in single or multiprocessor environments. One of the most promising features of Gem5 seems to be the simulation of unlimited numbers of processors, which can enable the exploration of memory subsystems and placement techniques for current and future server configurations. Of course, simulation performance may become a bottleneck for such kind of experiments, an issue that should get special attention in such a hypothetical future research.

Bit-Line Transitions in Energy Calculations The simulator can be improved to consider the effect of actual value transitions in the bit lines of the memory subsystem: If data values do not change with respect to the previous operation, no (dynamic) energy is consumed driving capacitive loads. Since the actual values of data exchanged with external memories may be different for each platform configuration, absolute energy consumption figures may also vary. The same considerations apply when calculating the energy consumed by the memory controller driving address and command lines. Even more importantly, some types of memories, such as resistive RAM (ReRAM), can avoid writing to bit cells that already contain the desired value, saving a respectable amount of energy—and increasing the cell's endurance.

Data values can be provided to the simulator modifying the profiling mechanism to include not only the address of each access, but also its value. At the expense of an increased log size, these values can be later supplied to the simulator. The simulation of SRAMs and DRAMs would require no further modifications because accesses are atomic and the simulator keeps the correspondence between addresses in the original execution and during simulation with the final placement decisions. However, cache memories would require more careful attention—probably storing actual data values in the simulated caches—because there is not a direct correlation between the dynamic data objects stored adjacently in a cache line during simulation and the original layout. An interesting remark is that integration with an architectural simulator such as proposed before would also provide the concrete data values exchanged with the external memories.

Analysis of Inactivity Periods Taking into account the inactivity periods of the memories would enable, for example, the exploration of energy-saving techniques. The designer could evaluate the effects of varying degrees of aggressiveness in moving memory modules into low energy modes and the impact on performance of reactivating them with more or less frequency.

In this last regard, an innovative proposal would be the exploitation of high-level information from the DMM to detect when a memory module does not hold any alive instances. As the system knows which pools are mapped into a given memory resource, it can poll the corresponding DMMs to check if there are any instances alive. If the module is not being used, it can be completely powered down without risk of data loss.

Even more, techniques such as presented by Lattner and Adve [17] might be used to migrate entire pools according to usage statistics and increase the size of the inactivity periods. At the cost of some access overhead—that should be evaluated and included in the trade-off—the system could migrate the pools and seamlessly transition into different placement solutions. Alternatively, if a memory management unit (MMU) is available, it could also be used to directly migrate whole pools with page granularity.

DMM During Simulation In the realm of dynamic memory manager simulation, the described implementation of our simulator uses idealized DMMs that create their data structures outside of the application data space and always use splitting and coalescing. This decision was taken to simplify the simulator implementation. In normal circumstances, DMMs build their data structures inside the pools themselves; thus, their own accesses may affect energy consumption in the platform to some degree. More research should be conducted to evaluate the impact of this decision or incorporate the actual DMMs during simulation.

However, it could actually be interesting to evaluate the effect of separating the DMM internal structures from the pool itself. Then, according to their number and pattern of accesses, they could be placed in different memory resources. Even if the memory footprint of the application grew, it might be possible to obtain interesting improvements due to more efficient DMM operations and the possibility of using more complex coalescing and splitting algorithms that are often avoided in DRAM memories to reduce the number of random accesses.

For example, in a 1 KB pool that serves exclusively requests of 24 B, the total number of blocks available for the application is 42. Depending on the space required by the internal DMM structures, their footprint might be 168 B (if 4 B per DMM block), 336 B (if 8 B per DMM block), or 504 B (if 12 B per DMM block). That space could be allocated from a tiny dedicated SRAM of 256 B or 512 B for very efficient DMM operations.

This approach has indeed been explored in the design of DMMs for hybrid memory subsystems that combine volatile and non-volatile memories. For example, WAlloc [30] separates metadata and data to limit the number of write operations to the NVM cells required to implement the management of memory blocks itself.

DRAM Bank Layout As explained in Sects. 5.5 and 5.6, throughout this work we assume that the address layout in DRAMs can be configured as "bank-row-column." This is possible in many systems, particularly in the case of FPGAs or ASICs, and is common in some DSPs. For general-purpose systems with caches, the usual layout tends to be "row-bank-column," while other options such as permutation-based interleaving [31] have been proposed at different times. Interestingly, recent AMD EPYC processors can change the interleaving configuration to suit the characteristics of different applications.

The address layout used in this text enables the placement of complete pools into DRAM banks ensuring that objects do not cross banks. However, other schemes favor long sequences of accesses (streams) because the row in the next bank can be activated in the background while a row in the current bank is accessed. Thus, big data objects (e.g., of many KB or even several MB) can be more efficiently accessed—in comparison, with the proposed layout 1 out of 512-to-2048 accesses could suffer a full row-activation delay, depending on DRAM row sizes. Although the advantages obtained by the ability to place pools directly on banks likely outnumber the possible inefficiencies encountered for long data transfers, future research might be conducted to evaluate the merits of these and other approaches.

DRAM Row Management We have not considered any policy to proactively close rows and thus we have not accounted for the potential energy savings. This is a topic worth future research that needs to be conducted at three different levels. First, policies for determining when to close a row. Second, DRAM banks consume less energy when no rows are open, but deeper "sleeping states" require more time before a row can be opened again. Third, the whole DRAM module can be pushed into energy-saving modes.

Regarding row policy, if the controller closes a row and it is accessed soon again, the new access incurs extra energy consumption and delay. As explained in Sect. 5.4.2, when DRAM accesses are sparse an interesting trade-off appears between closing a row and opening it again sometime later, or keeping it open all the time, because a bank with no open rows can enter a state with lower energy consumption. The extremes are usually known as "open-page" or "closed-page" policies. Some intermediate approaches try to identify groups of sequential accesses and close the active row proactively during the last one [9], a technique that is useful to at least partially hide precharge times since accesses to new rows will only wait for the row-activation time. The simulator does not implement any policy for proactively closing rows and thus does not currently account for the potential energy savings, but this capability can be easily implemented to conduct new research.

Additional energy savings may be obtained when none of the DRAM banks has open rows by carefully scheduling energy-saving modes for complete DRAM modules. The systems designed with our methodology may benefit from this possibility particularly because an efficient exploitation of SRAM memories may create long periods of time without any accesses to the DRAM; some special applications may even be run entirely on SRAMs. The simulator can be used to identify such conditions and explore new energy-saving schemes.

Pipelined Accesses Finally, the simulator may also be extended to implement pipelined accesses to SRAMs and caches, as multicycle accesses do currently produce stalls. In that way, more capable processor architectures could be evaluated.

7.2.1.7 Other Areas

Extended Applicability Our methodology works particularly well when several DDTs alternate footprint requirements (high resource recycling) or some groups are very accessed with interleaved sporadic accesses to other ones since it protects the most accessed instances against eviction. However, even when all DDTs have similar liveness and receive significant numbers of accesses, the methodology may still be interesting if spatial locality is low because movement of non-reused data words is avoided. Although some of the capacity could be underexploited because of the lower chances to reuse space through grouping, avoiding continuous (unproductive) data movements may produce significant energy savings. This effect may also apply for energy consumption in data-center servers.

Improving Cache Performance via DMM As we have explained throughout this text, DM often hinders the performance of cache memories. However, we envision two new promising techniques to exploit high-level knowledge from the DMM to improve the performance of cache memories.

First, the DMM knows when a memory range contains valid—in the sense of "alive"—data, whereas the cache controller does not: Normally, cache lines are marked as "valid" when they contain data copied from a further memory, and "modified" when these data have changed and must be written back before being substituted. However, when a dynamic data object is destroyed, its associated cache lines do not need to be copied back to main memory anymore. Similarly, when a dynamic object is created on an address range that does not currently reside in the cache, there is no advantage on copying any words to the cache because those words in main memory do not contain valid data—only cache-line allocation is required. This knowledge, if exploited efficiently, has the potential to significantly reduce the number of data movements across levels in a memory hierarchy. Furthermore, when a dynamic data object is destroyed, the corresponding cache lines (if any) may be marked as "invalid" to avoid copying them back to main memory in case of a future eviction. Indeed, in associative caches that line would be the first selected to store new data from main memory, saving the eviction of a potentially useful line in one of the other cache sets. All of this could result in interesting energy consumption and performance improvements.

Second, DDT knowledge obtained with our methodology might be used to improve the performance of cache-based hierarchical memory organizations. Inclusive cache hierarchies copy new data from main memory into all cache levels. When the new data are not going to be reused (e.g., during stream processing), the cache hierarchy suffers a phenomenon known as "cache pollution" that can severely reduce performance and increase energy consumption. Some processors support instructions to prefetch data directly into the L1 cache, so that the other levels are not polluted and the contents of the L1 cache can be recovered faster.

It may be possible to design a more efficient approach by marking the pages used by each pool with a maximum "level of cacheability" (e.g., in a "Page Attribute Table"). When data are moved closer to the processor, they would be copied only up to the maximum allowed level, thus avoiding interference with more important data. Even better, they could be allowed to reside in closer levels, but only if they use free lines without forcing any evictions—that situation would be enabled by the idea proposed in the previous paragraphs. The main drawback of this idea is that the processor architecture must allow direct word accesses to any of the cache levels. Nonetheless, it may be an interesting approach for ASIC or FPGA-based designs.

7.3 Applicability to Other Environments

The main aim of this work is to improve the cost of accessing dynamically allocated objects in embedded systems provided with SRAMs and DRAMs. However, as other computing systems become also more complex and their memory architectures less uniform, more relevant becomes data placement also for them, especially because in many cases the sheer amount of data and the complexity of the access patterns make the advantages of cache memories less clear. In this regard, an interesting experiment on the (un)suitability of complex cache hierarchies for scale-out workloads was presented by Ferdman et al. [11]. In spite of these observations, there is significant resistance to abandon the transparent mechanism of cache memories that so well has served us for many years. The following quotation summarizes the reasons why explicitly addressable memories have been relegated for a long time to the realm of embedded systems:

> One idea that periodically arises is the use of programmer-controlled scratchpad or other high-speed memories [. . .]. Such ideas have never made the mainstream for several reasons: First, they break the memory model by introducing address spaces with different behavior. Second, unlike compiler-based or programmer-based cache optimizations (such as prefetching), memory transformations with scratchpads must completely handle the remapping from main memory address space to the scratchpad address space. This makes such transformations more difficult and limited in applicability. In GPUs [. . .], where local scratchpad memories are heavily used, the burden for managing them currently falls on the programmer. [13, p. 131]

We may imagine the trade-off between energy efficiency and ease of design as a continuous between these two extremes. Computer architects have been pulling towards the side of ease of design for many decades, with cache memories offering an almost uniform view of the memory subsystem. Although the power and memory walls are pressing problems, favoring energy efficiency to the point of making the design of new systems unfathomable is neither an option. The lack of tools that help to tackle the complexity of the designs is probably the reason that inspires the past reluctance to adopt new mechanisms that may complement the cache memory. In this text we have seen how we can move a bit towards the other extreme, so that we recover some energy efficiency with bounded impact on complexity. The dynamic-memory based approach that we have shown does not require special care by the programmers: Minimum changes to augment the DM API—which might be introduced automatically by the compiler in the future—are enough. Dynamic data objects are then accessed as usual, via pointers (references). The reason is that this method does not mimic caching mechanisms to prefetch blocks of data before processing.

Nevertheless, the memory model has not been uniform for a long time. Consider, for example, the case of a dual-processor server with half its DRAM modules connected to the bus of each of the processors and an interconnection bus (e.g., AMD's HyperTransport) connecting both processors. The time it takes for each processor to access a memory word from a DRAM module depends on whether the module is connected to the processor local bus or it has to go through the interconnection bus; thus, the memory space is effectively non-uniform from a performance point of view. More modern architectures introduce these considerations even at the socket level; for example, AMD's Zen-based processors implement several "CPU complexes" (CCX) in the same die [1]. Several cores (e.g., four) integrate each CCX, each core with its own L1 and L2 caches. The cores in a CCX share an L3 cache. The DRAM memory channels are connected to different complexes, which means that the cores in a single socket have different latencies when accessing memory addresses on different DRAM modules.[2] In this type of architectures, the operating system has to be careful to consistently schedule each thread on the same NUMA group to avoid costly data movements. Therefore, it would seem that the time to study explicit placement techniques as a way to cope with the memory wall also in general-purpose computing systems has arrived.

Programmers can usually take advantage of the memory subsystem topology with specific functions to allocate memory local to a node (for example, with the use of GNU's libNUMA). But an automatic placement method that considers the characteristics of each element in this extended memory organization is still missing. Therefore, we propose our methodology for dynamic data placement as a starting point for this future work because it can be applied to any system in which the memory subsystem properties are a visible part of the programmer's model. As a natural choice, we further propose dynamic memory as the mechanism to implement data placement because it can adapt not only to the resources available when the execution starts, but also to the variations that happen during execution. Even more, the combination of dynamic memory and virtual memory may enable the reevaluation of placement decisions during execution.

The next paragraphs explain why the methodology that we have studied in this text is relevant for systems other than simple embedded devices outlining interesting research options in those areas.

7.3.1 Scale-Up Systems

The traditional model for improving system performance is using faster processors or more processors in a single system. The main characteristic of these systems is that the processors form part of a single system, usually with a single operating system. Quite frequently, these systems support a shared memory space in which every memory address is visible to every processor. However, to limit the complexity of the designs and to improve bus performance, memory modules and processors are clustered. Although all the memory modules employ the same technology, the cost of each access depends on the distance between the processor demanding

---

[2] AMD offers also an interleaving mode in which all the DRAM channels are accessed in parallel, thus having the same latency for all the memory addresses. This approach may improve bandwidth in streaming applications at the cost of increasing latency for other applications.

the data and the module containing them. Contention may also become a serious problem when several processors need to access the same memory module; thus, the interconnection network becomes critical. This model of computation is suited for the resolution of big individual problems.

The default approach of allocating memory in local resources and, when exhausted, from the closest ones, is becoming clearly suboptimal. Solutions for NUMA-aware memory allocation provide the mechanism to reserve space on concrete nodes (where nodes represent groups of processors and the memory modules directly connected to them). However, these APIs offer just the mechanism to allocate the space, often disregarding how each data object is going to be used: How that space is used is left for the application designer. Data placement can help to improve performance in these systems by ensuring that the most accessed data objects are located in the closest memory resources. Most importantly, by improving the tools presented in this work, the process may be executed automatically, freeing the designers from the burden of data placement across complex systems.

Multiprocessing adds complexity to the problem, signaling clear ways for research: For data shared among tasks running on several processors, should placement select a memory that is close to all of them (even if that node is not any of the ones involved with the data) or a memory that is close to one of them although the rest may pay higher access costs? In this regard, the work presented by Berger et al. [5], the Hoard dynamic memory manager, seems as an appropriate starting point to study the particularities of these systems and the modifications that should be incorporated into the placement techniques. Many opportunities remain in this area because solutions such as Hoard focus on efficient techniques for managing the pool of free blocks in multiprocessor systems, but they do not consider the characteristics of the underlaying memory resources.

7.3.2 Scale-Out Systems

A more economical alternative to big supercomputers is the scale-out model, where instead of creating more complex (and expensive) computers, a set of simpler ones is interconnected through regular networking technologies (e.g., TCP/IP over 10 Gigabit Ethernet) to create a big coordinated system. These systems are appropriate for problems that can be split into more or less independent parts or for tackling with swarms of simpler tasks that probably depend on vast distributed data repositories (e.g., NoSQL databases such as Apache Cassandra [2]). The scale-out model uses sophisticated algorithms to distribute the data aiming for improved performance and reliability against data losses.

Many modern data centers are built around the scale-out model. Their use cases include swarms of simple works (e.g., search queries), hosting of virtual machines from different users (e.g., scalable cloud computing), and complex jobs over huge data collections that exceed the storage capacity of any single node (e.g., "big data" problems). The last use case has led to the concept of Warehouse-Scale Computers

(WSC), where all the individual computers and the interconnection network are seen as a single big machine with particular characteristics.

In this new and exciting realm, we can foresee several situations where data placement solutions can benefit performance and, more importantly, reduce energy consumption. First, although applications are usually deployed in cloud services as complete virtual machines, the "OS-as-a-library" approach [18] may allow system programmers to build highly specialized solutions that exploit the underlaying memory resources. In a sense, these systems are an intermediate step between embedded systems and general NUMA systems where the complete set of applications is known at design time, keeping the same adaptability that is present in embedded systems. If the characteristics of the underlaying hardware are known to the designer, it can create adequate platform description files such as those used by $DynAsT$ and apply full optimization techniques such as presented in this work.

Second, some organizations are exploring the possibility of attacking the three walls at the same time changing the model of powerful and energy-hungry processors for a model composed of a sea of processing elements of modest performance that operate at a lower clock frequency and are attached to an intricate web of storage elements. This model can be suitable for the problem of big numbers of simple queries, where the relevant factor is not so much the complexity of the computations but their latency and avoiding "long tails" (jobs that take significantly longer to complete than the average) [28]. Clearly, data placement becomes a major issue in this paradigm. Distributed shared memory may also be an interesting target for data placement optimization.

Finally, the most interesting future development is probably the redesign of the rack in datacenters that companies such as Intel and Facebook propose. The likely advent of silicon photonics may open the door for new computing models where memory resources are separated and shared by many computing nodes. Applications should then choose carefully which data should be placed on local memory resources or on the shared (and presumably much bigger) pool. In some respects, that model would mimic the memory subsystem of many embedded systems, with the local DRAM playing the same role with respect to the shared memory pool than the SRAM plays with respect to the main DRAM. Therefore, a careful data placement might grant similar performance improvements and reductions in energy consumption. This may be particularly interesting in cloud environments where the infrastructure providers tend to supply machines with a low memory/cores ratios. If an application requires more RAM than processors, often the only alternative is to oversize the machine reservation with options that include both multiple cores and large memories, whose price tends to increase sharply. A promising alternative using the previous ideas could be using several normal machines connected to "memory servers." Interestingly, the access to remote memory with remote DMA over 10 Gb Ethernet might be significantly faster than the use of a local swap file in a magnetic drive and probably comparable to the bandwidth of a local flash solid-state drive (SSD)—more research on live shared environments would be required to evaluate this proposal.

7.3.3   New Memory Technologies

This text focuses on placing data objects on two types of memory technologies, DRAMs and SRAMs. However, during the last decade we have witnessed the emergence of new promising memory technologies. For example, new non-volatile memories (NVMs) promise persistence, low or moderate latency, high density, word addressability, and low energy consumption—particularly because of the absence of retention power, which allows embedded devices to power them off during idle times without state loss. On the other side of the spectrum, standard-cell memories (SCMs) can be used as a replacement for small SRAMs to improve energy consumption and reliability at low voltages [24], which can be a key enabler of near-threshold computing (NTC) [10].

7.3.3.1   Non-volatile Memories and the Fusion of Primary and Secondary Storage

Many types of NVMs compete today for widespread adoption: resistive RAM (ReRAM), magnetoresistive RAM (MRAM), phase-change RAM (PRAM), etc. However, their main drawback is their reduced write endurance in comparison to DRAMs or SRAMs. For example, preliminary academic works placed ReRAM endurance in the order of 10^{10} to 10^{12} writes per cell [27], whereas later characterization works suggested that it may be closer to 10^6 writes per cell [29] and commercially available chips are rated at 10^5 writes per byte [21]. These numbers suggest that ReRAMs are close to the endurance of flash-based storage, but too far from that required to substitute main memory. However, we can still think about using them with a careful data placement that, for example, places frequently written data objects in SRAM or DRAM, and objects that require persistence or that are mostly read in ReRAM.

Non-volatile memory technologies promise storage-class memories [12] with word addressability: The main differentiator with respect to, for example, flash-based SSDs or magnetic disks is the capability of accessing and updating individual memory words in time scales comparable to those of DRAMs or even SRAMs. Two possibilities appear immediately to exploit these new technologies. First, works such as Mnemosyne [25] and NV-Heaps [8] propose using NVMs to blur the distinction between primary and secondary storage, introducing the concept of persistent heaps. Their authors warn particularly about the need to prevent programming bugs, such as pointers in persistent storage that reference objects in volatile storage, because their effects would become permanent (i.e., not reverted by a system reset). This line of work can be very useful for big-scale applications such as distributed hash tables; an interesting overview of the promising field of in-memory computing is presented by Plattner and Zeier [22]. In this context, once the data objects are separated into volatile and persistent, techniques similar to the methodology presented in this text can be used to decide into which resource each dynamic data pool should be placed.

The second possibility consists on using any available storage technologies in a system as primary storage, not with the aim of persistence, but to increase the size of the working memory available to the processing elements to perform direct computation. Uniform algorithms, no need for serialization processes, and no transitions between user and kernel code for file accesses may reduce software size and increase application performance. In that line, HP Labs embarked several years ago in an innovative project named "The Machine" [15] and recently included in the broader scope of "memory-driven computing" [14]. If word-addressability is possible, then the only decision that remains is data placement. Once again, dynamic data placement is crucial to decide which data objects should be placed on each component of the memory subsystem, according to the characteristics of both.

In both cases, an adequate characterization of the behavior of the application's data types combined with careful placement can help to enjoy the benefits of these new memory technologies while palliating the negative effects of their limited endurance.

Even for block-oriented storage (e.g., magnetic hard disks), techniques similar to the ones presented throughout this text can help to break the traditional barriers between primary and secondary storage. As an example, let us consider the case of flash-based SSDs. At the time of writing this text, SSDs connected directly through the PCI-e bus (as opposed to via the disk controller interface) are common. Although the original idea of SSDs was to improve I/O performance that seems hardly to be the best we can do with them. For example, SDAlloc [3] offers a mechanism to use them as a repository of normal dynamic memory allocated with functions similar to malloc(). A careful management of dynamic memory and buffering allows SDAlloc to create the illusion of extended DRAM without blindly burning the flash storage. In comparison, placing the swap file of a regular non SSD-aware operating system on a flash disk may severely reduce its lifetime: Due to the way the DMM handles free memory blocks, it may be possible that physical memory pages get filled with a mixture of frequently and seldom accessed objects. Then, in order to serve application accesses, the operating system needs to move entire pages even if only a few bytes are going to be accessed—similar to the problem with cache memories that motivated this work.

Even with more classic technologies, we could envision ways of applying data placement to produce innovative solutions. For example, to use a magnetic disk as primary (working) storage, a simple first approach would use several DRAM buffers as working copies of the disk sectors, exactly in the same way that DRAM modules use row buffers to access complete cell rows—indeed, at a similar granularity because magnetic disks have sectors of 512 B or 4 KB, whereas DRAM rows commonly range in size from 512 words to 2048 words. A more involved approach for purpose-specific machines would implement sector management in a special controller directly connected to the processor bus, so that the processor could use regular load/store instructions with sector management performed transparently by the controller, exactly as the memory controller hides DRAM-idiosyncrasies from the processor. Once again, the key enabler is a careful data placement over all the available resources.

In a broad scope, the ideas for dynamic data placement presented in this work could enable the exploitation of multiple memory (and storage) technologies to improve system performance and reduce energy consumption. For example, SCMs and SRAMs may be reserved for frequently accessed collections of small data objects, DRAMs for bigger objects frequently accessed and modified (preferentially in sequential patterns), whereas NVMs would be reserved for mostly read or persistent objects. Flash storage would then be reserved for data accessed in streams and seldom written, and finally, a magnetic disk could be used to store collections of big data objects that are infrequently accessed but where a high proportion of these accesses are updates. Data placement would be the key to separate data objects with different characteristics and exploit the strengths of each memory/storage technology, generating a continuous view of the memory space in which the distinctions between primary and secondary storage blur.

7.3.4 Near-Threshold Computation and Inexact Computing

Near-threshold computing (NTC) is a design space where the supply voltage approaches the threshold voltage of the transistors [10]. NTC can bring substantial reductions in energy consumption at the cost of heavily reducing the maximum operating frequency. Logic elements can reliably withstand operation at near-threshold voltage (NTV) as long as the clock is reduced appropriately. However, conventional six-transistor (6T) SRAMs experience bit-flips when operating at sub-nominal voltages [4, 7], which can lead to errors in computations. For example, Bortolotti et al. [7] already found an error rate of 10^{-4} for a 6T SRAM when reducing the voltage just to 0.7 V (nominal voltage of 0.85 V in a 40 nm CMOS technology). Interestingly, many applications can support a certain amount of errors, either because they are inherently subject to noise in their inputs (and the errors introduced by the SRAMs at NTV can be seen as an additional source of noise) or because they produce qualitative results. This is the case of many applications in the biomedical domain [4] or of convolutional neural networks (CNNs). Often, a small number of errors will pass inadvertently to the user, as in the case of (low numbers of) erroneous pixels during video playback.

Inexact computing can also be applied in the case of embedded DRAMs (eDRAMs), for example, by reducing the refresh rate below the safety margins. Once more, the combination of modern frameworks to explore eDRAM error patterns (such as presented by Widmer et al. [26]) and placement techniques can enable new optimization opportunities to reduce energy consumption while retaining the required levels of correctness.

In both cases, knowing the characteristics of each of the application's data types and their tolerance to errors might allow the designer to produce data placements that take advantage of the energy reductions proposed by NTC while minimizing the effect on application accuracy.

References

1. Advanced Micro Devices: HPC Tuning Guide for AMD EPYC Processors. AMD, Santa Clara (2018). http://developer.amd.com/wp-content/resources/56420.pdf
2. Apache Cassandra (Last fetched on June 2019). http://cassandra.apache.org/
3. Badam, A., Pai, V.S.: SSDAlloc: hybrid SSD/RAM memory management made easy. In: Proceedings of the USENIX Symposium on Networked Systems Design and Implementation (NSDI), pp. 211–224. USENIX Association, Boston (2011). http://www.usenix.org/events/nsdi11/tech/full_papers/Badam.pdf
4. Basu, S.S.: Hardware/software co-design and reliability analysis of ultra-low power biomedical devices. Ph.D. thesis, Lausanne, Switzerland (2019)
5. Berger, E.D., McKinley, K.S., Blumofe, R.D., Wilson, P.R.: Hoard: a scalable memory allocator for multithreaded applications. ACM SIGPLAN Notices 35(11), 117–128 (2000). https://doi.org/10.1145/356989.357000
6. Binkert, N., Beckmann, B., Black, G., Reinhardt, S.K., Saidi, A., Basu, A., Hestness, J., Hower, D.R., Krishna, T., Sardashti, S., Sen, R., Sewell, K., Shoaib, M., Vaish, N., Hill, M.D., Wood, D.A.: The Gem5 simulator. ACM SIGARCH Comput. Archit. News 39(2), 1–7 (2011). https://doi.org/10.1145/2024716.2024718
7. Bortolotti, D., Mamaghanian, H., Bartolini, A., Ashouei, M., Stuijt, J., Atienza, D., Vandergheynst, P., Benini, L.: Approximate compressed sensing: ultra-low power biosignal processing via aggressive voltage scaling on a hybrid memory multi-core processor. In: IEEE/ACM International Symposium on Low Power Electronics and Design (ISLPED), pp. 45–50 (2014). https://doi.org/10.1145/2627369.2627629
8. Coburn, J., Caulfield, A.M., Akel, A., Grupp, L.M., Gupta, R.K., Jhala, R., Swanson, S.: NV-Heaps: making persistent objects fast and safe with next-generation, non-volatile memories. In: International Conference on Architectural Support for Programming Languages and Operating Systems (ASPLOS), pp. 105–118. ACM Press, Newport Beach (2011). https://doi.org/10.1145/1950365.1950380
9. Dodd, J.M.: Adaptive page management. US Patent 7,076,617 B2. Intel Corporation, 2006
10. Dreslinski, R.G., Wieckowski, M., Blaauw, D., Sylvester, D., Mudge, T.: Near-threshold computing: reclaiming Moore's law through energy efficient integrated circuits. Proc. IEEE 98(2), 253–266 (2010). https://doi.org/10.1109/JPROC.2009.2034764
11. Ferdman, M., Adileh, A., Kocberber, O., Volos, S., Alisafaee, M., Jevdjic, D., Kaynak, C., Popescu, A.D., Ailamaki, A., Falsafi, B.: Clearing the clouds: a study of emerging scale-out workloads on modern hardware. In: International Conference on Architectural Support for Programming Languages and Operating Systems (ASPLOS), pp. 37–48. ACM Press, London (2012). https://doi.org/10.1145/2150976.2150982
12. Freitas, R.F., Wilcke, W.W.: Storage-class memory: the next storage system technology. IBM J. Res. Dev. 52(4/5), 439–447 (2008). https://doi.org/10.1147/rd.524.0439
13. Hennessy, J.L., Patterson, D.A.: Computer Architecture: A Quantitative Approach, 5th edn. Morgan Kaufmann, San Francisco (2011)
14. HP Labs: Memory-driven computing: our vision for the future. https://www.labs.hpe.com/memory-driven-computing
15. HP Labs: The machine: a new kind of computer. http://www.hpl.hp.com/research/systems-research/themachine
16. Lattner, C., Adve, V.: LLVM: a compilation framework for lifelong program analysis and transformation. In: Proceedings of the International Symposium on Code Generation and Optimization (CGO), pp. 75–86. IEEE Computer Society Press, Silver Spring (2004). https://doi.org/10.1109/CGO.2004.1281665
17. Lattner, C., Adve, V.: Automatic pool allocation: improving performance by controlling data structure layout in the heap. In: Proceedings of the ACM SIGPLAN Conference on Programming Language Design and Implementation (PLDI), pp. 129–142. ACM Press, Chicago (2005). https://doi.org/10.1145/1065010.1065027

18. Madhavapeddy, A., Mortier, R., Rotsos, C., Scott, D., Singh, B., Gazagnaire, T., Smith, S., Hand, S., Crowcroft, J.: Unikernels: library operating systems for the cloud. In: International Conference on Architectural Support for Programming Languages and Operating Systems (ASPLOS), pp. 461–472. ACM Press, Houston (2013). https://doi.org/10.1145/2451116. 2451167
19. Marchal, P., Gómez, J.I., Piñuel, L., Bruni, D., Benini, L., Catthoor, F., Corporaal, H.: SDRAM-energy-aware memory allocation for dynamic multi-media applications on multi-processor platforms. In: Proceedings of Design, Automation and Test in Europe (DATE) (2003)
20. Marchal, P., Catthoor, F., Bruni, D., Benini, L., Gómez, J.I., Piñuel, L.: Integrated task scheduling and data assignment for SDRAMs in dynamic applications. IEEE Design Test Comput. **21**(5), 378–387 (2004). https://doi.org/10.1109/MDT.2004.66
21. Panasonic: ReRAM embedded super low-power consumption MCU MN101L (2014). https:// industrial.panasonic.com/ww/products/semiconductors/microcomputers/mn101l
22. Plattner, H., Zeier, A.: In-Memory Data Management, 2nd edn. Springer, Berlin (2012). https:// doi.org/10.1007/978-3-642-29575-1
23. Soto, M., Rossi, A., Sevaux, M.: A mathematical model and a metaheuristic approach for a memory allocation problem. J. Heuristics **18**(1), 149–167 (2012). https://doi.org/10.1007/ s10732-011-9165-3
24. Teman, A., Rossi, D., Meinerzhagen, P., Benini, L., Burg, A.: Power, area, and performance optimization of standard cell memory arrays through controlled placement. ACM Trans. Design Autom. Electron. Syst. **21**(4), 59:1–59:25 (2016). https://doi.org/10.1145/2890498
25. Volos, H., Tack, A.J., Swift, M.M.: Mnemosyne: lightweight persistent memory. In: International Conference on Architectural Support for Programming Languages and Operating Systems (ASPLOS), pp. 91–104. ACM Press, Newport Beach (2011). https://doi.org/10.1145/ 1950365.1950379
26. Widmer, M., Bonetti, A., Burg, A.: FPGA-based emulation of embedded DRAMs for statistical error resilience evaluation of approximate computing systems. In: Proceedings of the Design Automation Conference (DAC), pp. 36:1–36:6. ACM Press (2019). https://doi.org/10.1145/ 3316781.3317830
27. Wong, H.S.P., Lee, H., Yu, S., Chen, Y., Wu, Y., Chen, P., Lee, B., Chen, F.T., Tsai, M.: Metal–Oxide RRAM. Proc. IEEE **100**(6), 1951–1970 (2012). https://doi.org/10.1109/JPROC.2012. 2190369
28. Xu, Y., Musgrave, Z., Noble, B., Bailey, M.: Bobtail: avoiding long tails in the cloud. In: Proceedings of the USENIX Symposium on Networked Systems Design and Implementation (NSDI), pp. 329–341. USENIX Association, Berkeley (2013)
29. Yang-Scharlotta, J., Fazio, M., Amrbar, M., White, M., Sheldon, D.: Reliability characterization of a commercial TaOx-based ReRAM. In: IEEE International Integrated Reliability Workshop Final Report (IIRW), pp. 131–134 (2014). https://doi.org/10.1109/IIRW.2014. 7049528
30. Yu, S., Xiao, N., Deng, M., Xing, Y., Liu, F., Cai, Z., Chen, W.: WAlloc: an efficient wear-aware allocator for non-volatile main memory. In: IEEE International Performance Computing and Communications Conference (IPCCC), pp. 1–8 (2015). https://doi.org/10.1109/PCCC.2015. 7410326
31. Zhang, Z., Zhu, Z., Zhang, X.: A permutation-based page interleaving scheme to reduce row-buffer conflicts and exploit data locality. In: Proceedings of the Annual ACM/IEEE International Symposium on Microarchitecture (MICRO), pp. 32–41. ACM Press, Monterey (2000). https://doi.org/10.1145/360128.360134

Appendix A
Example Run of DynAsT

In this appendix we explore how a small application is instrumented, how the different steps of the methodology are performed with the help of $\mathcal{D}yn\mathcal{A}s\mathcal{T}$, and the output of the simulator for different platform configurations.

A.1 Example Application

The sample application is just a conceptual experiment with no real purpose. It has three data classes, v_1, v_2, and v_3. The first two are small data types of 500 B and 448 B, respectively, whereas the third one represents a class of big objects with a size of 1 MB.

The main code of the application creates an instance of v_3 that is alive during all the execution. This object is accessed in two bursts, with accesses to the other objects in between. One instance of v_1 and one of v_2 are created and destroyed consecutively, so that both objects are never alive at the same time (to enable an example of grouping). Both small objects receive many more accesses and thus, will be chosen by $\mathcal{D}yn\mathcal{A}s\mathcal{T}$ for placement sooner.

A.1.1 Source Code and Instrumentation

The declaration of the three data classes is the only part of the application affected by the exceptions-based instrumentation used in this example:

© Springer Nature Switzerland AG 2020
M. Peón Quirós et al., *Heterogeneous Memory Organizations in Embedded Systems*, https://doi.org/10.1007/978-3-030-37432-7

```cpp
#include "logged_allocated.hpp" // Instrumentation

class V1 : public logged_allocated<1>
{
  public:
  unsigned int data[125]; // 500 bytes
};

class V2 : public logged_allocated<2>
{
  public:
  unsigned int data[112]; // 448 bytes
};


class V3 : public logged_allocated<3>
{
  public:
  unsigned int data[262144]; // 1 MB
};
```

The rest of the source code is straightforward, with the only exception that the application's entry point cannot be main(), as this is defined by the profiling library itself:

```cpp
int MainCode(int , char **)
{
  V3 * v3 = new V3;
  for (int ii = 0; ii < 256*1024; ++ ii)
    v3->data[ii] = ii;

  V1 * v1 = new V1;
  for (int jj = 0; jj < 10; ++ jj) // Many accesses
    for (int ii = 0; ii < 125; ++ ii)
      v1->data[ii] = ii;
  delete v1;

  V2 * v2 = new V2;
  for (int jj = 0; jj < 10; ++ jj) // Many accesses
    for (int ii = 0; ii < 112; ++ ii)
      v2->data[ii] = ii;
  delete v2;

  volatile unsigned int aux = 0; // volatile prevents optimization
  for (int ii = 0; ii < 256*1024; ++ ii)
    aux += v3->data[ii];

  delete v3;
  return 0;
}
```

A.1.2 Instrumentation Output After Execution

The application is compiled normally and executed, creating the log file with the application memory allocations and data accesses. After execution, the profiling library outputs some useful information, such as the maximum application footprint, the number of allocations and deallocations (which should arguably be the same), and the number of reads and writes performed by the application on dynamic (instrumented) data objects:

```
$>SimpleTest.exe
Zeroing heap...DONE!
Starting address of heap: 4D30000

Maximum memory footprint: 1049076
Processed 526658 exceptions...
NumMallocs: 3 - NumFrees: 3
NumReads: 262144 - NumWrites: 264514
```

The log file for this example has a size of 2,633,368 B.

A.2 Processing with DynAsT

Once the log file is available, the designer can start working with $DynAsT$ in two different ways. The whole process, from analysis to simulation (or simply mapping), can be executed at once. However, the log files tend to be quite big for realistic applications (in the order of several gigabytes) and so it may be desirable to execute the analysis just once and save the results in a binary file for quick loading in later executions of the tool. This is particularly interesting because the analysis is always performed in the same way, independently of the design decisions that might be made in later phases.

Grouping may also be time consuming in some cases and thus, the same trick can be used with it. Of course, if the designer wants to change some of the grouping parameters, this step must be executed again. In this example, we show how the work with $DynAsT$ can be split into several steps through intermediate files.

A.2.1 Analysis

The analysis phase is easily executed with the following command line that instructs $DynAsT$ to dump the results of the analysis to the file "SimpleTest.analysis.bin":

```
$>dynast --InputLogFile log.bin --DoAnalysis
--AnalysisResults SimpleTest.analysis.txt
--DumpAnalysisFile SimpleTest.analysis.bin
--PrintAnalyzerStatsOnFPB
```

The dump file has a size of just 498 B. The following is an excerpt of the information written by the analysis step in the file "SimpleTest.analysis.txt":

```
Header is OK!
The log file is right and has 526664 packets.
Num of VAR_READ packets: 262144
Num of VAR_WRITE packets: 264514
Num of MALLOC_END packets: 3
Num of FREE_END packets: 3
Num of active blocks remaining: 0 (should be 0)
Number of distinct IDs: 3
Maximum memory footprint: 1049076
Max simultaneously active blocks: 2
Final memory footprint: 0


------------------
ID STATISTICS (order FPB > ID > Size)
------------------
ID: 1 SZ: 500      Read: 0        Writes: 1250    Created: 1 Dest: 1
MaxAct: 1 MaxFoot: 500      FPB: 2.50 Selfish: 0.99 SeqAcc: 1240
ID: 2 SZ: 448      Read: 0        Writes: 1120    Created: 1 Dest: 1
MaxAct: 1 MaxFoot: 448      FPB: 2.50 Selfish: 0.99 SeqAcc: 1110
ID: 3 SZ: 1048576 Read: 262144 Writes: 262144 Created: 1 Dest: 1
MaxAct: 1 MaxFoot: 1048576 FPB: 0.50 Selfish: 1.00 SeqAcc: 524286
------------------
```

DynAsT counts access operations when calculating the FPB, thus 1250 accesses on an object of 500 B yields an FPB of 2.50. The "Selfishness" and "SeqAcc" entries reflect DDT properties that may be exploited in the future.

A.2.2   Grouping

The grouping phase is executed with the following command line that instructs *DynAsT* to dump the results to the file "SimpleTest.grouping.bin":

```
$>dynast --LoadAnalysisFile SimpleTest.analysis.bin
--DoGrouping --GroupingResults SimpleTest.grouping.txt
--DumpGroupingFile SimpleTest.grouping.bin
```

The grouping dump file has a size of 766 B. The following is an excerpt of the information written by the grouping step in the file "SimpleTest.grouping.txt":

```
------------------
GROUP STATISTICS
------------------
Group 1 Read: 0 Writes: 2370 MaxFoot: 500 FPB: 4.74
ExpRatio: 0.63
ID: 1 SZ: 500 Read: 0 Writes: 1250 Created: 1 Dest: 1
MaxAct: 1 MaxFoot: 500 FPB: 2.50
ID: 2 SZ: 448 Read: 0 Writes: 1120 Created: 1 Dest: 1
MaxAct: 1 MaxFoot: 448 FPB: 2.50
Group 2 Read: 262144 Writes: 262144 MaxFoot: 1048576 FPB: 0.50
ExpRatio: 1.00
```

```
ID: 3 SZ: 1048576 Read: 262144 Writes: 262144 Created: 1 Dest: 1
MaxAct: 1 MaxFoot: 1048576 FPB: 0.50
-----------------
```

Thus, $\mathcal{D}yn\mathcal{A}s\mathcal{T}$ has correctly identified that instances of v_1 and v_2 have disjoint lifetimes and can be grouped together, whereas instances of v_3 should be placed apart.

A.2.3   *Mapping*

In its current version, $\mathcal{D}yn\mathcal{A}s\mathcal{T}$ converts automatically groups into pools and proceeds with the mapping step, which is the first platform-dependent step in the methodology. For this example, we have chosen three simple platforms that later allow showing the simulator capabilities. The first two platforms contain a small SRAM of 512 B and 128 MB of Mobile SDRAM or 256 MB of LPDDR2 SDRAM, respectively. The third platform has a 4 KB direct-mapped cache with 128 MB of Mobile SDRAM. $\mathcal{D}yn\mathcal{A}s\mathcal{T}$ is invoked with the following command lines for each of the cases, generating three mapping files of 258 B:

```
$>dynast --LoadGroupingFile SimpleTest.grouping.bin
--DoMapping --MappingResults SimpleTest.mappingSRAM_SDR.txt
--DumpMappingFile SimpleTest.mappingSRAM_SDR.bin
--PlatformDescriptionFile Plat_SRAM_512_LPSDRAM_128MB.txt

$>dynast --LoadGroupingFile SimpleTest.grouping.bin
--DoMapping --MappingResults SimpleTest.mappingSRAM_DDR2.txt
--DumpMappingFile SimpleTest.mappingSRAM_DDR2.bin
--PlatformDescriptionFile Plat_SRAM_512_LPDDR2S2_256MB.txt

$>dynast --LoadGroupingFile SimpleTest.grouping.bin
--DoMapping --MappingResults SimpleTest.mappingCache_D.txt
--DumpMappingFile SimpleTest.mappingCache_D.bin
--PlatformDescriptionFile Plat_Cache(d)4KB_LPSDRAM_128MB.txt
```

For the SRAM-based platforms the mapping report is the same. In both cases, one pool is placed on the SRAM and the other is placed on the DRAM:

```
-----------------
MAPPING STATISTICS
-----------------
There are 2 pools.

Pool 1 -  Reads: 0 Writes: 2370 MaxFoot: 500 FPB: 4.7
The pool has 2 IDs:
(1, 500) (2, 448)
The pool is split over 1 memory blocks:
Fragment 1: BlockID=0, Size=500, Address=0
------

Pool 2 -  Reads: 262144 Writes: 262144 MaxFoot: 1048576 FPB: 0.5
The pool has 1 IDs:
```

```
(3, 1048576)
The pool is split over 1 memory blocks:
Fragment 1: BlockID=1, Size=1048576, Address=0
------
```

The mapping report for the cache-based platform is slightly different. Here, both pools are placed on the main DRAM because the cache is transparent. Thus, the second pool is placed at offset 500 of the DRAM (BlockID = 0):

```
-----------------
MAPPING STATISTICS
-----------------
There are 2 pools.

Pool 1 -  Reads: 0 Writes: 2370 MaxFoot: 500 FPB: 4.7
The pool has 2 IDs:
(1, 500) (2, 448)
The pool is split over 1 memory blocks:
Fragment 1: BlockID=0, Size=500, Address=0
------

Pool 2 -  Reads: 262144 Writes: 262144 MaxFoot: 1048576 FPB: 0.5
The pool has 1 IDs:
(3, 1048576)
The pool is split over 1 memory blocks:
Fragment 1: BlockID=0, Size=1048576, Address=500
------
```

A.2.4   Simulation

The simulation is executed for each of the platforms with the following command lines:

```
$>dynast --InputLogFile log.bin
--LoadMappingFile SimpleTest.mappingSRAM_SDR.bin
--DoSimulation --SimulationResults SimpleTest.simSRAM_SDR.txt
--PlatformDescriptionFile Plat_SRAM_512_LPSDRAM128MB.txt

$>dynast --InputLogFile log.bin
--LoadMappingFile SimpleTest.mappingSRAM_DDR2.bin
--DoSimulation --SimulationResults SimpleTest.simSRAM_DDR2.txt
--PlatformDescriptionFile Plat_SRAM_512_LPDDR2S2_256MB.txt

$>dynast --InputLogFile log.bin
--LoadMappingFile SimpleTest.mappingCache_D.bin
--DoSimulation --SimulationResults SimpleTest.simCache_D.txt
--PlatformDescriptionFile Plat_Cache(d)4KB_LPSDRAM128MB.txt
```

A.2.4.1 Simulation for Platform with SRAM and Mobile SDRAM

The following text shows the output of $\mathcal{D}yn\mathcal{A}s\mathcal{T}$'s simulator for the first platform. In essence, the output is divided into four sections. First, the simulator gives the estimation of energy consumption and latency for accesses to the instances of each of the application DDTs. This information may be interesting to identify and evaluate specific algorithmic optimizations. Second, the simulator presents information for every memory module in the platform. In this case, "MemBlock 0" corresponds to the SRAM and "MemBlock 1" to the Mobile SDRAM. Next, the information on the interconnections is shown; however, these examples use a simple organization with a single bus and assign a null energy and latency cost for every transaction. Finally, the simulator prints the calculated total values of energy consumption and cycles in the memory subsystem.

```
-------------
ID STATISTICS
-------------
ID: 1 SZ: 500 Energy: 0.81 nJ Cycles: 1250
ID: 2 SZ: 448 Energy: 0.73 nJ Cycles: 1120
ID: 3 SZ: 1048576 Energy: 1284460.76 nJ Cycles: 4227032
-------------


----------------------
MEMORY BLOCK STATISTICS
----------------------
MemBlock: 0
Energy: 1.54 nJ Cycles: 2370 Reads: 0 Writes: 2370
McmBlock: 1
Energy: 1284518.41 nJ Cycles: 4227032 Reads: 262144 Writes: 262144
Page misses: 512
Empty cycles: 2370 Largest empty slot: 2370 cycles NumEmptySlots: 1
Avg. empty slot: 2370.00 cycles SlotsLongerThan1000 cycles: 1
EnergyReads: 640933.28 nJ EnergyWrites: 641284.63 nJ
EnergyActivations: 2242.86 nJ Background energy: 57.65 nJ
DelayReads: 2101248 DelayWrites: 2101232 DelayActivations: 24552
Total number of memory accesses to all modules: 526658
----------------------


----------------------------------
MEMORY INTERCONNECTION STATISTICS
----------------------------------
Interconnection: 1 Energy: 0.00 nJ Cycles: 0 Transfers: 526658
----------------------------------


TOTAL ENERGY CONSUMPTION: 1284519.95 nJ
TOTAL NUMBER OF CYCLES: 4229402
```

For the DRAMs, the simulator dumps extra information. For example, it identifies periods of inactivity and reports their number, average length, and longest one—in this example, the simulator identifies a single inactivity period between the two bursts of accesses to v_3 while the application accesses v_1 and v_2, which are placed on the SRAM. The simulator also reports the energy consumed during read, write, and row-change operations, and the background energy consumed by the DRAM being active but not accessed.

A.2.4.2   Simulation for Platform with SRAM and LPDDR2 SDRAM

The output for the second platform is very similar to the previous one, as they both have an SRAM and a DRAM. However, the simulator generates additional information specific to LPDDR2 memories. In particular, the number of accesses that correspond to the second word of each double-data-rate transfer—which are performed whether the processor accesses a single word or two consecutive ones—and the number of transitions between the logical states are reported:

```
-----------------------
MEMORY BLOCK STATISTICS
-----------------------
MemBlock: 0
Energy: 1.54 nJ Cycles: 2370 Reads: 0 Writes: 2370
MemBlock: 1
Energy: 288391.46 nJ Cycles: 1071036 Reads: 262144 Writes: 262144
Page misses: 256
Empty cycles: 2370 Largest empty slot: 2370 cycles NumEmptySlots: 1
Avg. empty slot: 2370.00 cycles SlotsLongerThan1000 cycles: 1
EnergyReads: 142074.70 nJ EnergyWrites: 145333.96 nJ
EnergyActivations: 917.04 nJ Background energy: 65.76 nJ
DelayReads: 528376 DelayWrites: 530396 DelayActivations: 12264
LPDDR2-S2 --> HiddenDDR: 262144 Idle2Read: 0 Idle2Write: 1
Read2Read: 262143 Read2Write: 0 Write2Read: 1 Write2Write: 262143
r2r_changerow: 127 r2r_samerow_hiddenddr: 131072
r2r_samerow_seamlessburst: 130944 r2r_samerow_fulldelay: 0
w2r_changerow: 1 w2r_samerow: 0
r2w_changerow: 0 r2w_samerow: 0
w2w_changerow: 127 w2w_samerow_hiddenddr: 131072
w2w_samerow_seamlessburst: 130944 w2w_samerow_fulldelay: 0
Extra tRAS delays: 0 Starting delays: 5588
Total number of memory accesses to all modules: 526658
-----------------------

TOTAL ENERGY CONSUMPTION: 288393.00 nJ
TOTAL NUMBER OF CYCLES: 1073406
```

A.2.4.3   Simulation for Platform with Cache and Mobile SDRAM

Finally, for the platform with a cache memory the simulator reports independently the energy consumption and latency for the caches and their associated DRAMs. The information about cache hits and misses is included so, although not used in this text, it could be employed in future analyses:

```
-----------------------
MEMORY BLOCK STATISTICS
-----------------------
MemBlock: 0
Energy: 2239943.49 nJ Cycles: 8654168 Reads: 524432 Writes: 262288
Page misses: 32786
Empty cycles: 559444 Largest empty slot: 2263 cycles
NumEmptySlots: 32777
Avg. empty slot: 17.07 cycles SlotsLongerThan1000 cycles: 1
EnergyReads: 1442014.23 nJ EnergyWrites: 640561.17 nJ
```

```
EnergyActivations: 143759.99 nJ Background energy: 13608.10 nJ
DelayReads: 4982160 DelayWrites: 2098304 DelayActivations: 1573704
CacheBlock: 0
Energy: 205447.81 nJ Cycles: 559444 Reads: 508048 Writes: 788946
Hits: 493881 Misses: 32777
Total number of memory accesses to all modules: 2083714
----------------------

TOTAL ENERGY CONSUMPTION: 2445391.29 nJ
TOTAL NUMBER OF CYCLES: 9213612
```

An interesting observation derived from this excerpt is that the use of a cache memory changes completely how the DRAM is used. Now, instead of a single period of inactivity of 2370 (CPU) cycles, the DRAM sees 32,777 slots with an average length of 17.07 (CPU) cycles (i.e., they are probably too short to allow the DRAM to enter an energy-saving state). In the future, we may devise further optimizations to exploit the better predictability of our solutions to transition the DRAM modules into energy-saving states more effectively.

Index

© Springer Nature Switzerland AG 2020
M. Peón Quirós et al., *Heterogeneous Memory Organizations
in Embedded Systems*, https://doi.org/10.1007/978-3-030-37432-7

Printed in the United States
By Bookmasters